Any Given Tuesday

Any Given Tuesday

A Political Love Story

Lis Smith

HARPER

An Imprint of HarperCollins*Publishers*

HarperCollins books may be purchased for educational, business, or sales promotional use. For information, please email the Special Markets Department at SPsales@harpercollins.com.

FIRST EDITION

Designed by Leah Carlson-Stanisic

Library of Congress Cataloging-in-Publication Data has been applied for.

ISBN 978-0-06-308439-1

22 23 24 25 26   LSC   10 9 8 7 6 5 4 3 2 1

In memory of my dad, Thomas Ramsaur Smith Jr.

Contents

Any Given Tuesday

Prologue

"Governor. *Stop*. It's *over*," the voice broke through on the conference call line.

Six months earlier, it would have been inconceivable that anyone, let alone a mere political consultant, would cut off the most high-profile, fearsome, and feared state chief executive in the country.

It was Andrew Cuomo who was talking, after all. He was eleven years into his reign at the top of the Empire State, and just one year removed from becoming a national phenomenon for his masterful, made-for-TV COVID briefings, which offered comfort to people amidst the isolation, confusion, and trauma of a global pandemic.

But on August 3, 2021—whether he was willing to accept it or not—he was a dead man walking. That morning, the Attorney General of New York released a bombshell report that concluded that he'd broken state law by sexually harassing women staffers in his administration.

"What's over?" Cuomo responded.

"*This*. All of *this*. *This* is over. There is *no* path forward for you," the adviser responded.

"It's *over* because I touched a woman on the back?" Cuomo shot back incredulously, his voice rising with a pitched tone of panic.

The adviser, someone not prone to hyperbole or challenging the governor unnecessarily, didn't mince words.

"It was more than touching a woman on the back. Don't bullshit yourself or us. If I, a man, were accused of doing any of the things you were, I would be out of a job by now."

Silence.

"So, you're telling me I don't fight back? I don't do a press conference? Why don't I just resign then?"

Silence.

"Lis," Cuomo started in his halting, Queens-inflected cadence, "what do *you* think?"

He was looking for a sympathetic voice, as he often did on calls. He had a knack for finding people who could agree with even his worst instincts. I paused before I answered.

It had taken me seventeen years—and twenty campaigns—to claw my way up the political ladder and go from a lowly field organizer to one of the top communications aides in the Democratic Party. Presidents? *Worked for one.* Senators and members of Congress? *Worked for a few.* Governors? *Worked for many.* Most recently, I'd served as a senior adviser on Pete Buttigieg's against-all-odds presidential campaign, where he'd defied conventional wisdom, won the Iowa caucuses, and become one of the Democratic Party's biggest stars. My star had risen as well.

I'd had an on-off professional relationship with Cuomo for the past three years, starting with his 2018 campaign, where I served as spokesperson and ran his debate prep. I reconnected with him at the beginning of COVID's onslaught on New York in March of 2020, when he'd call me for thoughts on his daily briefings. Now I was a part of his kitchen cabinet—the group of trusted, unofficial advisers—that he was relying on to help him weather the allegations of sexual harassment.

While Cuomo was notoriously tough on staff, he engendered a remarkable amount of loyalty in the people around him. Yes, he could be irrational and impetuous at times, but he matched that with a deep interpersonal warmth—showing up at weddings, bar/bat mitzvahs, and funerals. He always took the time to call staffers dealing with loss and personal hardships. He'd also been a formidable governor—the likes of which New York hadn't seen in decades. He'd managed to get the unruly state legislature under control and achieve some big things.

The last several months had tested that loyalty, as it became increasingly clear that Cuomo wasn't being straight with any of us—myself included. He'd led us down a path of defending him against claims of sexual harassment without giving us the full truth. We felt betrayed and misled.

"Governor, I'd like to disagree," I told him. "But I just don't see a way out of this."

In the moment, I meant it as much for him as for me.

ONE

First Crush

You always remember your first.

I started college less than a year after the Supreme Court hijacked the 2000 election. Their decision left half the country (myself included) extremely pissed off at the new commander in chief.

Ten months later and ten days before I'd set foot on Dartmouth's Hanover, New Hampshire, campus, I watched two different Boeing commercial airliners crash into the World Trade Center, reducing the Twin Towers to ash. For a brief moment, the anger over *Bush v. Gore* evaporated, and President George W. Bush rallied the American people to unite against the threat of global terrorism.

Like a lot of other kids in my generation—those two seemingly once-in-a-lifetime events (happening in the span of one year, no less) nudged me into the world of public service. It wasn't the hardest sell for me—ever since I'd seen *The War Room*—the behind-the-scenes documentary about the 1992 Clinton campaign, I'd had an itch to work in politics. The 2004 presidential race scratched that itch.

In a normal world, I'd have followed my parents' footsteps into the legal field, where my father had an especially distinguished career—and I still could. But nothing about the world felt "normal" anymore. The low stakes of "peace and prosperity" 1990s were gone—the politics of the 2000s felt like some life-and-death shit.

I settled on my 2004 candidate early: Senator John Edwards from North Carolina. I'd done some research and read a series of national profiles; Nicholas Lemann described him in the *New Yorker* as a "rollicking, full-throated, us-against-them populist." His "Two Americas" campaign theme drew me in immediately with its echoes of Bobby Kennedy.

Also, let's be real: his superficial appeal was an undeniable factor. He was youngish and vibrant, with the shiniest light brown hair you'd ever seen. Years of work as a courtroom star translated well on the Senate floor and campaign trail: he knew how to craft an argument, reel you in, and leave you with no other option but to side with him. With his pronounced Carolina drawl, he drew out his vowels irresistibly: "Eye-o-wah, ahhh bah-LEEEEVE in youuuu."

In March 2003, between my winter and spring terms at college, I dragged my parents down to DC to get an up-close look at "my" candidate. After he was elected to the Senate in 1998, Edwards's staff had developed a simple but effective political ploy: whenever the Senate was in session, he'd hold weekly, freewheeling sessions with constituents that his office dubbed "Tar Heel Thursdays."

Technically, I was cheating a little bit. I was a New York native enrolled at college in New Hampshire. But my dad's status as a born-and-bred Tar Heel got us in the door. Looking back, I'm a little surprised that my parents agreed to make the four-hour drive to Washington, but they were always supportive of the interests of their four children. And I think they were a little curious themselves. *Maybe this guy could be the one?*

It was a smallish crowd that day, thirty people tops. After Edwards gave a spiel about the legislation he was working on and answered questions, I made my pitch to him: "I want to run a student group for you at Dartmouth if you run for president." I pressed my email on his staff and got their cards.

A few weeks later, having joined forces with a couple dozen students and been elected co-president of Dartmouth Students for Edwards, I followed up with his staff to request a campus visit. I was annoyingly persistent. But here's the thing about being annoying (and persistent): sometimes it pays off. Less than a month after Edwards officially announced his presidential campaign in September 2003, he came to Hanover. I marked the occasion with my first ever op-ed in *The Dartmouth:* "Why Sen. Edwards Is the Answer." I wrote about his "fresh and innovative vision" and "uncanny ability to relate to common-folk America." It's hard not to cringe, looking back at my earnest twenty-year-old self. *Common-folk America? Really?*

I was a political naif at the time, so it felt like the biggest deal when his wife, Elizabeth, sent a kind thank-you email the morning the op-ed ran. We arranged for him to make a quick visit that evening to the Alpha Delta fraternity house—the basis for the movie *Animal House*, an edgy choice—followed by a more formal event at Dartmouth's Top of the Hop, a traditional stop for anyone running for president.

Thirty minutes before he was set to arrive on campus, I got a call from his staff. They were thrilled with how the main event was shaping up—the Hanover fire department was already turning people away from the door, meaning that it had hit the four-hundred-plus person capacity. But the AD trip was off; the campaign had decided that the optics of visiting a fraternity and especially *that fraternity* were not good. (They had a point. The last time someone had to be convinced to visit AD was when the *Animal House* character Boon told his reluctant date, "It's not gonna be an orgy! It's a toga party.") I have no idea what the behind-the-scenes negotiations entailed, but they somehow settled on a slightly absurd compromise—Edwards could give remarks *outside* the frat house, but he wouldn't be able to cross its rickety wooden threshold for the meet and greet we'd planned.

After delivering that slightly annoying news, the staffer threw in an additional wrinkle: "One other thing. We'd like you to introduce the senator at the Top of the Hop."

"Really? Are you sure?" I asked, silently panicking. "I didn't prepare anything."

"You'll do great."

To the Edwards staff, I seemed like an in-control, overconfident twenty-year-old. The chick who had bugged them endlessly about getting Edwards to campus and probably the type who would *kill* for the experience of introducing him. The caricature of an ambitious college Democrat.

The reality was a bit different. Ever since high school, when I was the lead violinist in the orchestra and a solo performer, I'd struggled with paralyzing stage fright. The second I was in front of a crowd, I'd start shaking or feel the uncontrollable urge to pee my pants. Embarrassing, I know.

Introducing a presidential candidate with twenty minutes' notice was the equivalent of sight-reading at a recital. Luckily, I still kept Propranolol—a beta-blocker I'd used for stage fright—on hand. I took twice what I normally would, but still found myself ducking behind a car on the main drag just minutes before Edwards pulled up. It was not the first or last time that I peed on a major thoroughfare around Dartmouth's campus, but it was the only time I did it stone-cold sober. I was fine by the time I had to introduce Edwards, delivering a serviceable, sincere speech, the type of remarks you'd expect from a college Dem warm-up act. When I handed over the mic, Edwards—ever the slick politician— nonetheless gushed to the crowd: "Isn't she great? Maybe she's the one who should be running for office here."

The high of that night was addicting. For years, I'd had my face smooshed up against the window of politics. After my first taste of what it was like on the inside, there was no way that I was turning back.

Dartmouth's schedule did me a solid. Unlike your normal college, it ran on a trimester system that required students to stay on campus the summer after their sophomore year. Sophomore Summer was one of those old-timey Dartmouth traditions, meant to build a sense of community among classmates. Mostly, it was just an excuse to enjoy the beautiful New Hampshire summer and drink Keystone Lights on the Connecticut River. It also freed up a term during junior year for us to get off campus and spend three months doing whatever we pleased.

I signed up to intern at the Legal Aid Society beginning in mid-January 2004. But first, in the days between Christmas and my start date in New York City, I'd be volunteering for Edwards in New Hampshire. The campaign put me up with a couple of other staffers at a picturesque farmhouse outside Keene—it looked like something out of a movie. Maybe a rom-com, maybe a slasher flick like *The Strangers*, but beautiful nonetheless. Every day, I'd brave the frigid New England winter to knock on doors and spread the word about the son of a mill worker who wanted to be the next president.

Fieldwork is where most people get their start for a very simple reason: it's the bitch work of campaigns. It's the political equivalent of being a telemarketer or a door-to-door salesman. And let's be real, *nobody* likes those people. Every day, you're more likely to get a door shut in your face than you are to find a sympathetic ear. It wasn't a great fit for me: I'm not a particularly gregarious person, and growing up, I was socially awkward to the point that I'd avoid picking up calls to my parents' house or even interacting with neighbors on the street. But like a lot of young women starting out in the professional world, I put enormous pressure on myself to deliver and to be *perfect*. A bad day at the doors or on the phones would make me feel like a failure. It's absurd to think about that now—the idea that somehow John Edwards's fate rested on my shoulders.

I've always been tough on myself, but it was also a sign of the times. Even though it was 2004, the gender politics of presidential campaigns felt like something out of the 1950s. There were few women in leadership positions. Men led the calls, ran the offices, and generally dominated the operations. I felt like I had to do double the work of my male colleagues to be accepted.

While "voter contact," as we called it, didn't come naturally to me, other things did. One morning, Edwards's New Hampshire deputy press secretary called: "Are you ready for some fun? We need someone to challenge Wes Clark on his record at his Peterborough town hall today." (It was a tactful way of saying that they wanted me to sandbag him, which in itself is an artful way to describe what is more commonly known in politics as "ratfucking"—essentially, taking extraordinary means to throw your opponent off their game.) Clark was a dark horse candidate, a late entrant into the race who was riding high on his apolitical role as a former NATO supreme allied commander and four-star general who had helped navigate an end to the complicated Kosovo War. He needed to be knocked down a peg or two.

I jumped in my car and drove forty-five minutes through the blinding snow with a mission: I—a junior in college—would trip up a guy who had negotiated directly with Slobodan Milošević to end the war in Bosnia and Herzegovina. The historic Peterborough town hall—another structure that looked like something straight off a movie set—was buzzing when I walked in. It was packed to the gills with nine hundred or so Granite State voters (and some out-of-state political tourists—a common sight right before the primaries) and their very, very puffy outerwear. I listened to Clark's stump speech and then sprang into action once the event coordinator called for questions. I started out on the ground level, raising my hand gamely—no luck. I ran up to the balcony on the second floor—still no luck. *Shit, shit, shit.* I went back to my original perch and held up my hand once again. Clark finally pointed in my direction, and the mic found its way to me.

"General Clark, I respect your record of service and your views on money in politics," I began. He smiled, taking in the compliment in front of the warm crowd. "But I would like to hear how you square that with your record of lobbying for big corporations and some of the most brutal dictators—how can we trust you?"

A frown replaced the smile. He was *pissed.* He knew he'd been had. When I walked out of the venue, the incognito Edwards tracker—campaign-speak for the usually fresh-out-of-college staffer tasked with the unenviable job of following around the opposition and capturing their every utterance—no matter how mundane—on camera raised his chin at me from across the room. Clark's top communications aide, whom I recognized from TV, called out as I headed to my car: "Tell them you did a good job." He didn't know which campaign I was with, but he knew that I didn't fit the profile of an interested Peterborough voter.

I made the drive back to Keene buzzing with adrenaline. I blasted music and drummed the steering wheel of my Nissan Xterra, which by then was filled with trash from the trail—fast-food wrappers and empty cans of soda, cups of gas-station coffee, and packs of

Marlboro Lights. The odds of being picked to ask a question at a nine-hundred-person town hall are low, but through persistence, I'd succeeded. The moment clearly stuck with General Clark; years later when addressing the College Democrats National Convention in St. Louis, a friend in attendance told me that he'd joked about the time a rival campaign had sent an unassuming, "cute college girl" to one of his events to pose a gut punch of a question.

As my start date at Legal Aid approached, I felt a pit grow in my stomach. I loved the high of working on a campaign, and I wanted to see this one through: the idea of packing up and leaving midstream to push paper for lawyers filled me with dread. Thankfully, when I emailed Legal Aid to let them know I'd be sticking with the presidential race, they wrote back that they understood and wished me luck. My dad, who was on the board of Legal Aid, was also surprisingly cool about the last-minute decision: "I'm just happy that you've found something that you love doing," he assured me. He'd always been tough to win over, so his approval was all I needed.

The timing couldn't have been better. Edwards's candidacy—previously an afterthought in a field crowded with known quantities like Senator John Kerry, Governor Howard Dean, Speaker of the House Dick Gephardt, former vice presidential candidate Joe Lieberman, and Clark—started to take off. Dean and Gephardt were engaged in a murder-suicide in Iowa, exchanging increasingly nasty attacks that turned off voters. It created an open lane for a fresh-faced outsider like Edwards to rise. On January 11, he received the coveted *Des Moines Register* endorsement—the biggest boost any candidate can receive prior to the Iowa caucuses. Eight days later, he finished a close second to John Kerry. It seemed like a foregone conclusion to the campaign that Edwards was the story of the night. I went to bed with the giddiness I'd felt when I'd lost my first incisor, placed it under my pillow, and waited for the tooth fairy's visit. Instead of a dollar under my pillow, I expected to find glowing national coverage of Edwards when I opened my eyes in the morning.

Just as I'd eventually learned that the tooth fairy wasn't real, I learned the morning after the Iowa caucuses that the news wasn't always, either. The media narrative wasn't centered on the Kerry-Edwards result in Iowa. It was singularly focused on Howard Dean's speech from the night before. And not in a positive way.

On first blush, Dean's fiery, rallying-the-troops entreaty to a beer-fueled, rowdy crowd of more than three thousand supporters in Des Moines hardly seemed newsworthy. As was the case with any candidate who had worked his or her tail off in Iowa, Dean's voice was taxed that night. He downplayed his third-place finish but grew increasingly animated as he worked up to his conclusion:

> *Not only are we going to New Hampshire. We're going to South Carolina! And Oklahoma! And Arizona! And North Dakota! And New Mexico! We're going to California! And Texas! And New York! And we're going to South Dakota! And Oregon! And Washington and Michigan! And then we're going to Washington, DC, to take back the White House.* YeeeeAHHHHHH!

The speech seemed unremarkable in the moment—a little spirited, sure, but it was the night of the Iowa caucuses, for God's sake. By the next day, though, you couldn't turn on your TV without reliving Dean's "scream." It was the first viral moment in American politics, and it hit before the advent of YouTube, Facebook, Twitter. To the reporters in the room where Dean spoke, there was nothing off. But it was a different story for TV executives in New York and Washington, DC. The audio that was piped back to them was distorted by Dean's unidirectional microphone—meaning it picked up Dean's audio, but completely drowned out the cheers of his supporters. He sounded completely unhinged. Thus, the "Dean Scream" was born. He never recovered.

Dean wasn't the only candidate who was screwed by the media's obsession with his speech. In a normal world, Edwards would have been the big day-two story coming out of Iowa. For months, no one had taken him seriously, yet he'd somehow surged to win 32 percent of the vote. Still, his finish was relegated to an afterthought because it didn't make "good copy" quite the way Dean's speech did. It was the first, but not the last, time I saw a candidate with a surprising finish in Iowa robbed of their glory. (More on this later.)

I've thought about that moment a lot over the years, especially after I got into the communications side of politics. There were plenty of fair criticisms to lodge against Dean: he was a bit of a loose cannon; his campaign had squandered a big lead and blown through millions of dollars and sent into Iowa a weird army of people in orange hats who turned off locals. (An aide for a rival campaign had famously compared the crowds that showed up to Dean events to "the bar scene from *Star Wars*.") Dean's speech that night, however, was not one of his missteps. Why, then, was his scream such a big deal?

For better or for worse (for much worse, in my estimation), it foreshadowed the media's addiction to "gaffe" coverage—easily digestible gotcha moments that guarantee eyeballs, usually at the expense of more complex issues. Dean's scream made for better TV than talking about the actual results of the Iowa caucuses . . . or delving into the complicated issues that could or should define an election like this one: the Bush administration's refusal to pay heed to intelligence that al-Qaeda was planning an attack on US soil in the lead-up to 9/11 or their hasty march to war with Iraq or their broken promise to deliver tax relief to low-income Americans (in reality, the average millionaire received a tax break of $120,000, while the bottom 20 percent of workers got a lousy $27 back from the government under the Bush tax cuts). The news business was cultivating and catering to the appetite of its consumers: more ice cream, less spinach.

Edwards didn't get the predicted bump out of Iowa. And we couldn't lay it entirely at the media's feet. There was no appetite for his brand of politics in New Hampshire. His honey-dipped drawl and smooth-verging-on-slick pitch fell flat with voters— it's called the Granite State for a few reasons, not least because its residents are pretty steely. I realized we were in trouble when my Franconia-based grandparents broke the news that they were voting for Dean. If I couldn't win over my own family, how could we win over everyone else? Edwards got smoked—he came in fourth place, earning just 12 percent of the vote. Making it worse—for me, at least—was that Clark had edged him out for third place— outpacing him by just nine hundred votes.

I spent the next month driving up and down the East Coast and Midwest. I went to South Carolina, where Edwards won easily, thanks to his roots in the state. Then Wisconsin, where he finished a close second to Kerry, breathing new life into his campaign. And finally Ohio, a Super Tuesday showdown state. It may have been my first political campaign, but I quickly learned to smell the stench of death that wafts over a campaign right before the end. It hit Edwards in Toledo. Just three hundred fifty people showed up for his speech at the local university—the last couple rows of seats were empty. It was an abrupt change from the usually packed rooms that greeted him. When I glanced over at the press riser, I saw the beat reporters who'd been covering him for months exchange knowing looks with one another: the end was upon us.

The next day—Super Tuesday—Kerry won every single state in

contention. As we were absorbing the shellacking at our makeshift office in Toledo, we got word that Edwards would be dropping out of the race the following day in Raleigh, North Carolina. So we hopped in our cars and drove through the night to catch his concession speech.

Huddled with the other staffers, I sobbed uncontrollably. Edwards was my first political love. When his campaign ended, it hurt as badly as my first breakup. Was Edwards the perfect candidate? No, but I was a young, impressionable college student who had dropped everything and traveled the country for him. The experience had been more transformative than any course I'd taken in school; I'd seen the nitty-gritty of political campaigns up close, I'd found my calling.

Here's the thing that people outside of politics might not get: it's never just a job for the people who work in it. It's a passion project—one that's deeply personal and hopelessly emotional.

Losing sucked, but learning about the "real" John Edwards years later sucked more. I didn't get involved in his 2008 campaign, yet I was rooting for him to win the nomination nonetheless. If I were honest with myself, the Edwards of 2008 didn't have quite the same spark. When I'd see him at events, the "aw-shucks" authenticity that had roped me in initially registered as tired, almost like an act. *We get it. You're the son of a millworker.* He never took his candidacy to the next level that he would've needed to win.

Even worse, the *Huffington Post* published an extremely carefully worded story about his professional involvement with a New Age videographer who'd followed him around the country between 2005 and 2008. The article—"Edwards Mystery: Innocuous Videos Suddenly Shrouded in Secrecy"—noted the large payments that his political action committee had made to her, and how every single one of her videos had been erased from the web. The subtext was clear: there was some funny business going on. (Somewhat ironically, the author of the story was Sam Stein—a Dartmouth classmate of mine who'd serve as an elected officer of the Students for Edwards group. While Sam had made the leap from college partisan to independent journalist, I doubted that it gave him pleasure to write a story admitting that he too had been duped by Edwards.)

The chatter remained largely in the background until Edwards dropped out of the race. Within months, it would explode when the *National Enquirer* chased down Edwards at a Beverly Hills hotel during a rendezvous with the videographer and her infant daughter, whom he'd secretly fathered. Talk about going out with a bang.

There he was—a former presidential candidate golden boy, VP nominee, husband to a publicly adored wife with Stage IV breast cancer—hiding in a hotel men's bathroom to avoid having his photo taken and his secret life exposed. Everyone who had worked for him was disgusted.

By this point I'd been around the block a couple of times, but the Edwards scandal was still a real wakeup call. Like any sentient human being, I'd watched Bill Clinton bob and weave his way through the never-ending allegations of his extramarital affairs. I'd just hoped that Edwards was different.

He wasn't, and it was a stark reminder about the nature of people who run for political office. Even the most noble politicians possess a healthy streak of narcissism. Public adulation is intoxicating; it's easy to get sucked in by the trappings of power—the rooms filled with important, rich people hanging on their every word, the private planes, the groupies and sycophants who enable and encourage their bad behavior, and the naive true believers who carry their water from one campaign stop to the next.

Beneath the glossy, charming exteriors, politicians are still flawed human beings, and if they seem too good to be true, they usually are.

The "Two Americas" Edwards pitched on the campaign trail wasn't the disadvantaged vs. the advantaged. It was the people who bought his bullshit and the people who saw right through it. Sadly, I fell in the former camp.

I wouldn't be fooled again, I told myself.

Yeah, right.

TWO

The It Factor

On August 30, 2005, Missouri state auditor Claire McCaskill announced that she would be challenging Republican Jim Talent for the US Senate seat from the Show-Me State. It received little mention in the national press.

For starters, Talent was the guy who'd carried out the difficult feat in 2002 of running against and defeating Jean Carnahan, who'd gotten elected to the US Senate under the most bizarre of circumstances. In 2000, her husband, Mel—a former two-term governor and the then–Democratic nominee for the US Senate— died in a tragic plane crash three weeks before the election. In death, he still managed to defeat John Ashcroft, becoming the first person in history elected posthumously to the US Senate. Jean was appointed to a two-year term to fill the seat. In the definitive ranking of races that are tough to run, let alone win, taking on a sympathetic widow is near the top. At the top? Trying—as Claire was—to beat the guy who'd beaten the widow.

Then, of course, there was the category 5 hurricane that had begun pummeling the Gulf Coast of Louisiana the day before Claire's announcement. Little did anyone know at the time, but a few winds that had picked up steam over the southeastern Bahamas would—four hundred forty-two days later—help Democrats sweep the midterm elections, giving them control of the House of Representatives, and sending five new blue senators to Washington, including one Claire Conner McCaskill from Houston, Missouri.

President George W. Bush was taking his fateful flight over New Orleans's collapsed levees on the day I arrived in St. Louis to begin working for Claire. I was with my boyfriend Jeff Smith who, one year earlier, had become a minor local political celebrity when he'd come within 1 percent of winning the Democratic primary to replace the outgoing congressman Dick Gephardt. The winner of Jeff's race came from an impeccable political pedigree. His mom had been a US senator, his dad a governor: his last name was Carnahan.

Now, you might be wondering how I found myself just barely west of the Mississippi River, moving in with a former congressional candidate (who happened to share my less famous last name). That's a longer story.

The winter term I'd spent working for Edwards had spilled over into a summer and fall spent working for Tom Daschle as a field organizer in South Dakota. I lived in Madison, a small town fifty miles north of Sioux Falls, in an apartment over a bakery. The baker was my landlord. Every morning, I'd buy his chocolate glazed doughnuts before I made the twenty-five-foot walk across the street to the campaign office I shared with another twentysomething New Yorker on the campaign. For four months, I worked from nine A.M. to nine P.M. every day, with the sole goal of winning over voters in the local rural community. I recruited a small army of politically eager high school boys to help me knock on doors and deliver yard signs. Those boys and I screamed our asses off in the Moody County courthouse when we learned on election night that we'd beaten our vote goal by twenty-three whole votes. Sometimes the smallest victories in life feel like the biggest ones.

Even so, Daschle lost, and it was time for me to return to school and face reality. It seemed like an enormous comedown; I'd lived a serious, adult life—or so I thought—and now I had to go back to the world of my sheltered classmates. The prospect of dorm life was beyond dull.

I drove up to Hanover the night before classes started. It was late when I started to move my stuff into my room, and after I got a suitcase and some sheets up the two flights of stairs, I collapsed: I was sick as a dog. It was as if my body was manifesting my internal anxiety at returning to school. I loved Dartmouth—I would still have good times there—but I wanted to be in the real world with real people, not back on a college campus.

I slept through my first class, a senior seminar for which a 100 percent attendance rate for the term was mandatory. *Oops.* I would probably have slept all day if my friend Nina hadn't busted through my door around noon. "Lis, I just met your soul mate," she announced breathlessly. "You need to transfer into my ten A.M. class. You will love this professor, and he will love you. You guys are the same person."

In a feverish haze, I informed her that I couldn't: My dance card was full in terms of classes. I was already pushing Dartmouth's limits with the number of terms I'd taken off. Still, I asked his name, and when she replied, I realized that the professor in question would be teaching my afternoon class that day. "Okay, we'll see," I told her, just hoping to get her off my back.

I limped across the barren—but always beautiful—Dartmouth green in my South Dakota–acquired cowboy boots and took a seat in the back row of the class. I didn't expect much. I thought Nina was being hyperbolic. And then Jeff waltzed into the room.

He was late. Stressed out. Slightly disheveled, shirtsleeves rolled up unevenly. A few inches shorter than I was. He had a cold and a runny nose that he kept wiping with the back of his hand. But he was *hot*. He was *magnetic*. Even before he opened his mouth, you could see stars forming in the eyes of every female student in the room.

At Dartmouth, we were used to crusty old teachers. The professors in their forties looked like they were in their sixties. The professors in their sixties looked like they were in their eighties. They were gruff and New England-y. They had prematurely gray hair and wore orthopedic, snow-worthy footwear to class. They'd sometimes learn our names, but—by and large—weren't especially engaging. This guy was *different*.

Jeff bounced around the room frenetically as he did icebreakers with each of the students about their favorite upcoming race, sometimes delivering sarcastic comments in response. This was *not* what we were used to. When he got to me, I told him my pick: the New Jersey gubernatorial election where Senator Jon Corzine would likely be the nominee. It was a niche choice that threw him off guard. "Huh. Are you from New Jersey?" *Nope.* "Have you worked in politics before?" *Yup.* I thought I saw a flash of something in his eyes, a flicker of connection.

Somehow, over the course of ninety minutes, I'd gone from completely too cool for school to completely infatuated with my professor. It could've been the fever talking—I headed from that class straight to the infirmary, where I was held for two nights with a 104-degree temperature.

Imagined connection notwithstanding, my crush on Jeff appeared to be completely unrequited. Fourteen minutes into the (nonmandatory) office hours session I carefully dressed for, I caught him looking at his watch; exactly one minute later he ushered me out and welcomed in the nerdy male student waiting in the hall. The line for Jeff's office hours were the longest I'd ever seen at Dartmouth.

Yet slowly but surely we started corresponding over email. I'd always been a little boy crazy, and during my personal rumspringa off from Dartmouth I'd dated and crushed on a lot of guys, almost always older. I knew this wasn't just a passing thing, though, when I found myself more excited by an email notification from Jeff than a text from some cute New York investment banker.

The rules about professor-student relationships at Dartmouth were hazy; there was no policy against students and professors dating (they'd clearly internalized the New Hampshire state motto of "Live Free or Die"). Every year, rumors flew around campus about students and their professors—usually it was some twenty-year-old involved with their fiftysomething married art professor. But it was always in the shadows, gossip that people tittered about, nothing that was ever truly corroborated.

Jeff and I were different. When we started seeing each other that spring, he was just thirty-one years old. I was twenty-two. My parents had met him during the break between terms, and both signed off on the relationship. We didn't try to hide anything; we were boyfriend and girlfriend. We were in love—a tawdry *Lifetime* movie, this was not.

Still, it caused a kerfuffle on campus. *The Dartmouth*—the official campus newspaper—wrote a pointed story about the tension of students and professors dating. Jeff was called in by the head of the government department and asked whether he thought it was appropriate to have "an affair" with a student. "My God," he deadpanned, "is she married?" I totally get why colleges have bans on professor-student relations, but Dartmouth didn't. If it were such an affront to them, they probably should have codified it somewhere.

At the end of the spring term, Jeff told me that he wanted to run for office again in St. Louis.

I spent the summer after graduation shacking up with him in his college-provided housing. By night, we plotted out the early strategy for his state senate run; by day, I studied for the LSATs, certain that law school was my next step. *No one serious works in politics*, I told myself, even as I paused my studies to make fundraising calls to Jeff's donors from Dartmouth's library café.

Jeff also wanted me to move with him to St. Louis—a tall order. I'd never been to the city, except the one time I'd toured Washington University during my college visits. The thought of relocating there gave me hives, but I was *in love*. I probably would've moved to Papua New Guinea with Jeff, if he'd floated the idea. But one thing working in Missouri's favor—unlike Papua New Guinea— was that it was home to a marquee 2006 US Senate showdown. Before I took the leap and agreed to move with him, Jeff connected me with a top staffer for McCaskill. After a couple of phone interviews, I was hired for an entry-level communications job. So, St. Louis it was.

I was one of Claire's first Senate campaign staffers, and I threw myself into the role enthusiastically. In the first few months, it was a bare-bones crew, and I took on whatever assignment they handed me—everything from fundraising calls to transcribing hours and hours of Claire's media interviews from over the years to drafting emails, press releases, and letters on her behalf.

I got my first taste of the electoral challenges that lay ahead when we went to Jeff's parents' house for dinner. For obvious reasons, the usual "how did you meet" icebreaker questions were off the table. Jeff's parents were a tad more conventional than mine.

The conversation quickly turned to what was next for me as a St. Louis transplant.

When I told his mom that I'd started a job on Claire's campaign, she responded: "Well, good luck. I like Talent. He's moderate. He's one of the good ones." And that, in the form of a passive-aggressive put-down from a potential future mother-in-law, was Claire's biggest challenge.

She wasn't running against a villainous goon—Talent seemed benign enough. *He'd beaten a widow, for God's sake.* We struggled with this fact on the campaign, settling on messaging that conceded Talent was a nice guy, just not someone on the side of Missourians. At an early staff meeting, we brainstormed ways that we could dent Talent's favorability with voters. Our campaign manager half-jokingly offered a bounty to any staffer who could capture Talent's alleged secret cigarette habit. That's how little material we had to work with. Still, 2006 was shaping up to be a bad year for Republicans, and Talent would soon drop the Mr. Nice Guy act.

It was the first time I'd had the chance to spend a substantial amount of time around a candidate I was working for. And there was just something about Claire. Back then I didn't know what to call it. Smart? Brassy? Willful? Now I do—she, somewhat improbably, had an "it" factor about her.

Sure, she was a little overweight. She overdid the black liner and mascara (something our ad consultant got her to drop mid-campaign). She didn't pretend to be glamorous and didn't make that a part of her appeal. But she was a natural-born communicator: observing her in action, it was clear why she'd been such a formidable prosecutor earlier in her career.

She could turn almost every argument thrown at her on its head—in town halls, press conferences, interviews, or debates. She was the queen of pivots and straight talk, even if she wasn't always talking straight. (This was politics, after all.)

In real time, I watched her demolish preconceptions about how a woman candidate should carry herself:

"Too ambitious?" You bet I am. We need ambition on our side in Washington.

"Unladylike?" I grew up in Houston, Missouri; I didn't take waltzing lessons there.

"Too blunt?" Where I'm from, we tell the truth, whether or not people wanna hear it.

"Unlikeable?" If there aren't people who don't like me, then I'm not doing my job. Trust me, people will hate me in Washington.

It would do Claire disrespect to say that it was effortless. Even as she masterfully navigated lines of attack and deflected tough questions, you could almost feel her wrestling them to the ground and making them cry uncle. Sure, she knew how to use a scalpel, but her weapon of choice was a hammer.

At times, she could come across as a little too glib. When she'd land a killer line—prepared or not—her face would set into a shit-eating grin. She was good, and she knew it.

In other words, she embodied all the traits of a successful male candidate. She projected strength, unapologetically. She was who she was.

That's a long-winded way of saying: Claire did not give a fuck.

Wherever that strength came from, she was going to need it for that campaign.

In December 2005, when I was in Washington, DC, for an Edwards reunion party, Jeff called me to deliver horrific news: Claire's first husband, the father of her three children, had been found shot and killed in a car in Kansas City. To this day, no one has ever been charged with the crime. It was devastating for Claire and her children. A different candidate might have let it derail his or her campaign. Claire did not.

That summer, our director of compliance died in a skydiving plane crash. She was an outgoing young woman who stocked her office with mini candy bars because she loved visitors. Three others on board the plane died along with her; it was a complete freak accident that shook our close-knit team. We held a memorial service for her family at a park veranda near our suburban St. Louis office, during which Claire gave a eulogy that had the entire audience in tears.

A couple days after the service, an important campaign finance disclosure was due. We gave notice that the paperwork would be late, thinking it more than understandable given that the person responsible for filing it had just died tragically. The Missouri Republican Party accused Claire of exploiting "the unfortunate death of a staffer to hide from her incompetence." It was like salt in the wounds for everyone on the campaign, seeing our deceased coworker's name invoked so casually to score a cheap political point.

When it rains, it pours. Within weeks, Fox News was calling us and threatening to run an interview with Claire's current husband's first wife, during which she made explosive allegations of physical abuse. Fox held the video over our heads like an anvil. *You know we have this on tape. We can run it at any time.* For weeks, we went back and forth with them—doing our best to persuade them not to run the story. We talked to the reporter, the producer, and even Fox News executives about how irresponsible it would be to air it—an allegation like that could end a campaign. We produced signed affidavits from Claire's husband's children disputing the charges. After evaluating the evidence, Fox backed down—they weren't going to touch it. We breathed a sigh of relief, but we weren't naive. Whenever a Fox reporter or anchor would call asking for access to Claire, there was always a veiled reference to *the thing* and the promise: "I won't go there."

A few years later, I ran into one of the people behind the interview at a state fair—he wanted the candidate I was working for to do an interview with Fox. He leaned into me and whispered conspiratorially: "Remember, we did you a solid with that Claire story."

Whatever journalistic restraint Fox had in 2006, they lost by 2018, when they ran with the headline: McCaskill's Husband Was Accused of Abuse by Ex-Wife. When the story was inevitably converted into a TV ad by the Club for Growth on behalf of Claire's opponent, Josh Hawley—now more commonly known as that guy with overly gelled hair who fist-pumped the air in solidarity with the Trump dead-enders and QAnon supporters shortly before they stormed the House and Senate on January 6, 2021—her husband's ex-wife released a statement calling it "terribly unfair and the worst kind of disgusting dirty politics."

In October, Claire and Talent faced off in a *Meet the Press* debate moderated by Tim Russert. It was unquestionably the biggest national moment for our campaign; we prepared for it like it was the Olympics. Russert was the toughest interviewer in TV news, and we knew that he'd eat his Wheaties that morning.

And did he ever. He pounced on Talent with his opening question, grilling him on how House Republicans had covered up allegations that Florida representative Mark Foley had sent explicit photos and messages to congressional pages as young as fifteen.

It was a full-blown scandal for the moralistic GOP. Talent struggled through evasive answer after evasive answer, while Claire was direct in calling for the House Republican leadership to resign. Claire 1, Talent 0.

But Russert wasn't going to go easy on Claire. He trained his fire on her own political associations, asking about Bill Clinton's upcoming campaign swing on her behalf. It was a politically sensitive topic—she couldn't win in a red state like Missouri without putting daylight between herself and national Democrats. The upside of having the support of Clinton, a prolific fundraiser, was matched by the clear downside of being tied to a polarizing, former Democratic president who had been impeached for lying about sexual misconduct.

> MR. RUSSERT: *You're having Bill Clinton come in to raise money for you. Do you think Bill Clinton was a great president?*

> Ms. McCaskill: *I do. I think—I have a lot of problems with some of his, his, his personal issues. I said at . . .*

> MR. RUSSERT: *But do you . . .*

> Ms. McCaskill: *I said at the time, "I think he's been a great leader, but I don't want my daughter near him."*

In the campaign war room back in St. Louis, the comment met with a chorus of whoas and holy shits. It was true to form for Claire—she was loath to pull a punch. But still, she'd chosen to take on the last Democratic president—without question, the dominant figure in her own party—on *Meet the Press*, no less. Today, in the post-#MeToo era, Claire's comments would seem mainstream, but back in 2006, they were positively radical.

There was an extra wrinkle to the whole affair, no pun intended. The next day Claire was scheduled to be in New York City for a slate of big-dollar fundraisers. The primary organizers behind them? The Clintons, a pair not exactly known for their magnanimity toward their many critics. One by one, they pulled down each of the events they'd sponsored, leaving Claire twisting in the wind. Back in St. Louis, we were getting hourly updates about how the New York trip was devolving.

I got a more personal view into all of it. For candidates campaigning in New York, it's not practical to walk or take public transit everywhere. It's also a big waste of money to hire a car service to sit outside meetings and events to do nothing but circle the block and burn gas. Smart campaigns hire volunteer drivers to squire candidates around town. After some trial and error, we'd settled on an impeccable chauffeur, a New York native and devoted Democrat with a personal stake in Claire's campaign: Adrienne Sullivan Smith, aka my mom. She called me in disbelief after she dropped Claire off that day. She told me how she'd had to console Claire—someone you didn't exactly associate with vulnerability—as she'd wept quietly in the passenger seat of my mom's yellow Mini Cooper. (A truly ridiculous choice for transporting a potential US senator, but I digress.)

Is it understandable that the Clintons were pissed at Claire's comments? Sure, even politicians are allowed to have human emotions. But it wasn't like Claire was the first Democrat to call out Bill Clinton's behavior, and it wasn't like she had any other option. What was she going to do? Go on *Meet the Press* and say it was totally kosher for Clinton to receive a blow job from a twenty-two-year-old intern? That would've gone over *great* with the people of Missouri.

The Clintons weren't exactly political virgins; on some level, they had to understand the predicament Claire faced. Their outsized response offered a window into how they operated—their egos came first and took precedence over electing Democrats who crossed them. Ultimately, the canceled events cost Claire six figures of fundraising dollars, hardly an insignificant sum in a race where she was neck and neck in both fundraising and the polls.

The Clintons may have gotten their revenge that day, but they did not get the last laugh. Claire won the election, of course. But she did one better: she got even.

During the 2006 campaign, she began to forge a close relationship with then–US senator Barack Obama. Even though he was less than two years into his Senate term, he was arguably becoming the most electric campaigner in the party—nearly as in demand as President Clinton. He didn't have the same decades-long juice with donors, sure, but you could feel change afoot in the Democratic Party. When he campaigned for Claire at the Uptown Theater in Kansas City, the two-thousand-plus supporters in the room screamed when he came out onto the stage, *dancing*, no less, to Stevie Wonder. I caught myself cheering like a high school girl at a Backstreet Boys concert. He returned repeatedly to Missouri to campaign for Claire, including the weekend before the election.

The events of that fall—the Clinton betrayal, Obama's dogged campaigning—all came full circle in January 2008, when Claire became the first female US senator to endorse Obama in his face-off against Hillary Clinton for the Democratic presidential nomination. It was a *major* coup for his campaign and seen as nothing short of a slap in the face to Clinton, the first serious female presidential candidate, who was banking on support from women in her campaign.

Claire didn't just issue a press release, like many senators do, and leave it there. No, she took a page from the Michael Jordan playbook: it had become "personal" for her. She was on TV, she was on the campaign trail, she was firing back right and left on Obama's behalf as the Clinton campaign leaned increasingly into accusing Obama—and his campaign—of sexism. (He certainly didn't do himself any favors on that front in the debate where he famously patronized her with the line, "You're likable enough, Hillary." *Oof.*)

Now, there is no doubt in my mind that Claire had genuine respect and affection for Obama. You could feel it even in 2006 when, behind the scenes, she'd confide in us: "I think he could be president. You just watch." But still, I don't think she ever, for a second, forgot the way the Clintons had treated her. It was a lesson in the karmic nature of politics—today's pissant state auditor could be tomorrow's formidable US senator.

In October, Michael J. Fox came to St. Louis to campaign for Claire and the Amendment 2 ballot initiative, which, if passed, would legalize stem cell research in Missouri. Missouri had a thriving medical research sector, with large hospitals like Barnes-Jewish in St. Louis and nationally renowned institutions like Washington University, so the measure had clear economic implications for the state. Stem cell research was also seen as having the potential to unlock cures to diseases like Parkinson's and Alzheimer's. Fox had become the most visible face of Parkinson's in America after he'd gone public with his diagnosis in 1998.

The debate presented Talent with a no-win dilemma. While the country club, business-friendly wing of the party—led by the Chamber of Commerce, sitting Republican governor Matt Blunt, and former US senator John Danforth—supported the measure, the evangelical, right-wing base—led by Missouri Right to Life—vehemently opposed it. Talent cast his lot with the latter group, but he did his best to avoid the topic in interviews and on the campaign trail.

I drew the straw of staffing Fox and preparing him for his media interviews. Up close, I was shocked by the severity of his Parkinson's symptoms; the Michael J. Fox I saw that day had the rakish demeanor of his iconic *Back to the Future* character Marty McFly, but physically, it was a different story. When his interviews began, I found myself wiping away tears as I watched him struggle to subdue his tremors, sweating visibly and contorting his entire body to hold down his shaking legs.

After his first interview—with Jo Mannies, the *St. Louis Post-Dispatch*'s lead political reporter—ended, he turned to me and asked how he'd done. Politics wasn't his area of expertise, and he was clearly a bit nervous.

"You were great!" I told him. "The points you made were really, really powerful."

He paused for a second, then responded with a wink, "Well, *naturally.*" We broke out in laughter, and his comically timed faux bravado cut the tension in the small room.

Prior to that interaction, Parkinson's had seemed so foreign to me, something tragic that affected other people. I didn't have a crystal ball that could predict how—seven years later—I'd get a call that my dad was diagnosed with it. Or how I'd once again run into Michael J. Fox when he was campaigning for Pete Buttigieg and get to thank him for his strength and sense of humor and tell him how his example had helped me navigate the personal devastation I felt watching my dad struggle with the disease. Or how fifteen years later, I'd sit at my dad's bedside as he passed away from complications tied to it. Or even, in the more near term, how seismic an effect Fox's involvement would have on Claire's race.

It's the rare national campaign where celebrities don't try to insert themselves. Ours was the rare example where it made a positive difference. When Fox was in town for that round of interviews, he cut an ad for Claire.

You didn't need to be a political savant to see how explosive it would be. We posted it online on a Friday evening—to little fanfare—but saved its television debut for the next night during Game 1 of the World Series, when Missourians would be glued to their TVs watching the St. Louis Cardinals vie for the title against the Detroit Tigers.

It featured none of the typical bells and whistles associated with political ads—it was just Michael J. Fox straight to camera. "As you might know, I care deeply about stem cell research," Fox said in the ad's opening shot. "In Missouri, you can elect Claire McCaskill, who shares my hope for cures. Unfortunately, Senator Jim Talent opposes expanding stem cell research. Senator Talent even wanted to criminalize the science that gives us a chance for hope. They say all politics is local, but that's not always the case. What you do in Missouri matters to millions of Americans—Americans like me."

The words themselves were powerful, but the image even more so. Fox had largely receded from his acting career in the years prior. The visual of the still boyish American icon lurching across the screen, struggling with the debilitating symptoms of an incurable disease—was jarring. There was no music, no magic camerawork, no cutaways. It was as stark a political ad as I could remember seeing.

Overnight, it became the most talked-about ad of the 2006 cycle. It exploded online and on cable news—Claire's fundraising soared. Hundreds of thousands of dollars in the first twenty-four hours. Millions within a week. A thirty-second TV ad undermined Talent's core strength with voters—that he was a reasonable, decent, and empathetic Republican. No one except the most hardened Republican partisan could watch it and believe that Talent was as inoffensive as his campaign had portrayed him.

National Republicans tried to fight back with a response spot featuring their own roster of celebrities—actors like Patricia Heaton, the mom in *Everybody Loves Raymond*, and Jim Caviezel, the star of Mel Gibson's *The Passion of Christ*—to warn voters that the stem cell ballot initiative would "make cloning a constitutional right." On the Monday after we launched the ad, Rush Limbaugh—the Darth Vader of right-wing radio—did his best to dismiss it to his 10 million listeners during his daily radio show:

> *He is exaggerating the effects of the disease . . . He's moving all around and shaking and it's purely an act . . . This is really shameless of Michael J. Fox. Either he didn't take his medication or he's acting . . . This is the only time I've ever seen Michael J. Fox portray any of the symptoms of the disease he has. He can barely control himself.*

For dramatic effect, Limbaugh mimicked Fox's tremors for the cameras that were livestreaming his show to the web, all the while chuckling to himself like a schoolyard bully. The backlash was instantaneous. It only drove more donors to our campaign and more eyeballs to the ad on YouTube.

It was a bridge too far, even for a piece-of-garbage provocateur like Limbaugh. A guy who'd made his name by never apologizing had to backtrack and eat shit on air for mocking an American icon with Parkinson's. His comments dominated the news coverage in the final weeks of the campaign, and they helped turn Claire's campaign into a cause célèbre for national Democrats. Naturally, for every action, there's an equal and opposite reaction, and it energized the socially conservative opponents of Amendment 2—illustrating how there's no such thing as a "clean win" in American politics.

Talent himself was clearly feeling the heat of Claire's challenge. During one debate prep session in the early fall, Claire interrupted a mock debate session after the stand-in for Talent delivered a nasty personal quip about Claire's husband's business dealings. "Cut the crap, guys," Claire told us. "I know you're trying to get me prepared, but Jim's not going to do that. He's not that type of guy." Weeks later, she was proven wrong when Talent spent the majority of the fourth Senate debate attacking her husband's finances, accusing him of being a tax cheat. She had been half right, I guess. Talent delivered the clearly prepared attack lines with all the comfort of a guy undergoing a colonoscopy on live TV. But whatever personal decency he still possessed clearly had taken a backseat to his desire to win at all costs.

The never-ending nature of the attacks against Claire was exhausting and exacerbated by the dysfunction playing out behind the scenes. A well-oiled machine her campaign was not. The national Democratic Senatorial Campaign Committee—then run by Chuck Schumer—had to send in top officials to take the helm.

The drama took its toll on our staff. A coworker—a kind, low-drama guy—who sat at a desk within earshot of mine went to the ER with symptoms of what he thought was a heart problem. After wearing a monitor for a few days, he was informed that—thankfully—he didn't have any cardiac issues. He was just suffering from extreme anxiety. Within days of each other, another campaign staffer and I were diagnosed with shingles, a virus usually associated with senior citizens and immunocompromised people, not healthy twentysomethings. One day as I was leaving work, I searched the office frantically for my car keys. Unable to find them, I traced my steps back to the parking lot, figuring I may have dropped them along the way. What I discovered, instead, was that my car was still in its spot. The keys were still in the ignition. And the engine was still running—ten hours after I'd parked it.

And yet the long, brutal campaign paid off. Shortly after midnight on November 8, 2006, Claire declared victory in the Senate race; she defeated Talent by a margin of forty-nine thousand votes out of over 2 million cast, finishing with 49.6 percent of the vote to his 47.3 percent. It was an impressive feat, especially considering the dysfunction and drama behind the scenes—a flawed campaign operation, the murder of her first husband, personal attacks against her second husband, threats from the right-wing noise machine, and seeming efforts to hurt her candidacy from the two most prominent Democrats in the country. Even in the best of years, with a perfect campaign, it's an uphill battle to beat an incumbent who hasn't committed a fireable offense. Through sheer will, grit, and an increasingly favorable political environment, Claire managed to pull it off.

Her campaign was a crash course in crisis communications, specifically a critical rule: compartmentalization. She never let the opposition see her sweat, and by extension that revealed a side to me that I never knew existed. I'm a naturally anxious and jumpy person, but I never felt calmer than in the heat of a crisis. I was *in control*. I could *handle* it. I was addicted to the drama and the rush I'd feel as a crisis bubbled up.

My crisis communications skills would become my trademark. But later I'd learn that even I had my limits. My overconfidence in my abilities led me to make dumb decisions while facing crises of my own. My go-to attitude was "I alone can fix this," even when I clearly couldn't.

Over the years, I'd find myself suppressing and rationalizing inconvenient facts that contradicted the narratives I desperately wanted to believe. I isolated myself personally. I created emotional lockboxes for realities that were too painful to confront and tried to talk myself through them clinically and mitigate them as if they were just another PR crisis. If I could control other people's narratives, surely I could control my own?

Still, I'll never forget that night when Claire went from a Senate challenger to a US senator-elect. Little did I know it would be the last victory party I'd attend for six long years.

THREE

The Life of the Party

The craziness wasn't limited to my work life. The excitement of my whirlwind courtship with Jeff in beautiful Hanover soon met the reality of living in St. Louis—a city I had no affinity for and where I knew literally no one. Our social lives revolved around the bleak local political events and dinner parties where his friends could barely hide their disdain for our professor-student relationship. I, in turn—with all the maturity of a twenty-two-year-old—did not hide the disdain I felt for their parochial lifestyles and worldviews. Jeff called me a snob; I called him weak (but in stronger terms). It was not a recipe for a healthy relationship.

Two months into living in St. Louis, I leveled with him: "I can't do this anymore." I was absolutely miserable. I started to go through the motions of moving back to New York. But I realized I didn't want to leave Claire's campaign, and I didn't want to leave Jeff either. I wanted a boyfriend who treated me like an equal and made sure his friends did as well. I also wanted a life of my own—

one that didn't revolve around his. To his credit, Jeff made a promise to be more attentive and responsive, and he kept it.

Maybe because we were finally honest with each other, things began to fall into place. I developed a core group of friends my age—three St. Louis investment bankers who became my best friends. Every Friday and Saturday night, we'd hit the town, meeting up at restaurants or bars or lounges, often ending up at the all-hours strip clubs across the river in Illinois. During the week, we'd get together after work, pre-game with cheap liquor, and go see midlevel bands like Wilco, Death Cab for Cutie, and My Morning Jacket play at the local music venues. A couple times, we scored tickets to the once-a-month, always-sold-out shows that Chuck Berry would still put on at Blueberry Hill—a local burger joint with an intimate downstairs concert space. I was much happier living out the social life of someone my age than pretending to be a political spouse.

Not that the life of a political "significant other" was all that glamorous. Jeff was an adjunct professor, and I had an essentially entry-level campaign job, so neither of us was rolling around in dough. We lived downtown next to the Botanical Gardens in a shabby, $500/month apartment furnished with hand-me-downs from his parents and yard sale giveaways. I remember seeing him lug in the most god-awful, worn-down sofa: "Can you believe I got this for free?" *Yes. Yes, I could.* We never got cable for the little TV that sat on a cardboard box in our "living room." A year or so into our lease, I splurged on bunny ears so I could watch Roger Federer play in the Wimbledon finals.

Neither of us cared about interior decor, and neither of us was particularly neat. There were things that *mattered*—our love for each other, our families, our ambition, our careers. The state of our living quarters was the last thing on our minds. Time hasn't changed my perspective on things. My apartment in the West Village of New York is entirely furnished with my dad's hand-me-downs from when he lived in Washington, DC. Of all the things I'd spend money on, furniture would be last on the list.

Our limits were tested at times. There was a two-month period when we had a brutal infestation of mice. We tried to take it in stride. One day, we walked into our apartment and saw one lying dead on the floor—bad sign. We could hear them scurrying around at all hours of the day and night—*especially* at night. We bought glue traps instead of the traditional ones, and I developed my own little routine to keep things humane. The second I'd hear a mouse squeaking on a trap, I'd put on oven mitts, grab a bottle of vegetable oil, and walk a couple blocks down from our apartment to liberate the unlucky critter. It would take a few glugs of oil and some careful nudges as the usually terrified mouse nipped at my mitt, but I'm proud to report that there were never any mouse fatalities in the Smith household (that lone, initial fallen soldier aside). We finally met our breaking point the night when I woke up to a mouse scurrying across my pillow and hair. That was a bridge too far. Jeff called the exterminator the next day.

Jeff's campaign, though on a more local level, was as crazy and intense as Claire's. He was a white, Jewish guy running in a field of eclectic candidates in a majority Black district. Every day, he had to deal with the unique insanity of local city politics.

He could never fully shift out of campaign mode. He insisted on writing personalized thank-you notes to every single one of his campaign donors, whether they'd donated $3 or $1,000. Those days, with email and digital staffers—even for a state senate campaign—it was pretty much unheard of. It was impractical and extremely time intensive. Most nights, as I slept quietly by his side, Jeff would be up until four or five A.M. grinding out notes in his chicken scratch handwriting. I assumed that at some point the stack of envelopes he brought home would dwindle, but they kept getting bigger and bigger. It seemed like utter madness: he was campaigning sixteen hours a day and getting by on just a few hours of sleep. But there was a method to his madness: the reason why the stack kept getting bigger and bigger was because of his personal touch—people gave again and again. It helped him raise an ungodly amount in low-dollar donations for a local election.

It didn't take long for our personal relationship to get dragged into the campaign. During the spring of 2006, a prominent Republican consultant who ran a gossipy Missouri blog decided to train his attention on me. At the time, personal attacks against candidates and candidates' family members were relatively commonplace. But it was pretty rare for a young campaign staffer to be on the receiving end.

The consultant trawled my high school friends' Facebook pages for incriminating information. He posted a photo of one of my friends in a bikini, alleging it was me and questioning my judgment—as if wearing a two-piece were a crime. His focus got creepier from there. He found a satirical Dartmouth Facebook group that labeled me the "sex symbol" of the "Dartmouth Communist Party"—a nonexistent entity, as anyone familiar with Dartmouth campus politics would know—and had reporters call our campaign to ask for comment about whether or not I was a card-carrying member of the Communist Party. *Seriously?* He "broke" the news of my live-in relationship with Jeff, sensationalizing the nature of how we'd met at Dartmouth and tried to present it as some kind of tabloid scandal. It felt like every week there was something new.

After a few of these particularly invasive posts, Claire invited me out for margaritas near the campaign office to blow off some steam. "Fuck him," she said, referring to the physically large strategist who was targeting me. "You know he writes this stuff about you because he hasn't seen his dick in years." Claire could've been making calls to donors, schmoozing at some big-money event, or even spending time at home with her family. Instead, she took ninety minutes out of her afternoon to give me a feminist pep talk in a near-empty Tex-Mex restaurant.

The attacks from the right coincided with attacks that Jeff was starting to face from within the race. One of his opponents, a white, socially conservative former elected official from South St. Louis, ran a coded campaign implying that Jeff—a slight, unmarried, thirty-three-year-old candidate—was gay. He put out a mailer featuring a photo from Jeff's first St. Louis fundraiser, in which he was holding his blazer over his shoulder. The photo was doctored so that you couldn't see the blazer, only what looked like a flamboyant hand gesture. The mail piece screamed, "JEFF SMITH ISN'T BEING STRAIGHT WITH YOU"—not exactly a dog whistle. More like a full-throated scream: "GAY!" Jeff also had to deal with anti-Semitic attacks from hatemongers like Louis Farrakhan, who campaigned for one of his opponents and labeled Jeff a "colonizer" for daring to run as a Jewish man in a majority Black district.

A few weeks before Jeff's primary, a St. Louis political blog wrote up a "push poll" that was going around the district. Recipients were asked if they'd be "more or less likely to vote for a candidate if you knew he was paying somebody to pretend to be his girlfriend to cover up the fact that he is gay?" Jeff took it in stride, quipping to reporters who asked about it, "Don't worry, I couldn't afford her." My skin was a bit thinner, and the attacks wore me down.

We went into Election Day with no feel for Jeff's chances. There was no way to credibly poll a state senate district like that, and it was a total crapshoot with two white candidates and three Black candidates in a district that was racially, ideologically, and economically diverse. So we were shocked when the race was called early for Jeff that night. He'd beaten the nearest competitor by 11 percent. Winning the primary was as good as winning the general, given the overwhelming Democratic nature of the district. Jeff was on his way to Jefferson City.

We'd been through a lot as a couple. In 2005, meeting cute as professor-student, falling in love, and moving to his hometown together. In 2006, enduring simultaneous, stressful campaigns. In 2007, trying to further establish our professional careers—him as a star in the Missouri legislature, me as an aspiring, hungry communications staffer bouncing from one gig to the next. In February 2007, I moved to Kentucky to work as a traveling press secretary for a Democratic gubernatorial candidate. (I was hired by the Edwards New Hampshire press secretary who'd sent me to the Wes Clark town hall.) In June, after my Kentucky candidate lost, I moved back to Missouri and helped engineer a Republican state senator's party switch to the Democratic Party—he'd go on to become the attorney general of the Show-Me State. In December, I moved to a suburb of Chicago to work for a top-tier Democratic congressional candidate.

Welcome to the crazy world of politics. You move state to state. You uproot your life, test your personal relationships, get your face pounded into the dirt, and still get up the next day and go to work. It's not a world for people who crave stability, the people who don't believe in taking risks, or the people who think that the peak of human existence is a nine-to-five job with nice benefits and quiet weekends. The allure of the political game is the draw of any given Tuesday—the belief that for every loss, there's a winning candidate or cause or campaign that redeems your faith in the process and creates the positive change you've been seeking all along.

The downside, of course, is that it's nearly impossible to lead a normal, healthy, balanced life. It was tough enough to maintain my relationship with Jeff when we were living in the same state, but even harder when I was living in three different states during the course of one year. So it probably shouldn't have come as a surprise when he called me a few days after the November 2008 election— one that my candidate lost: "I don't think this is working out anymore," he told me. "We just haven't gotten along well in a long time. It probably doesn't make sense for you to come back here."

My first reaction was shock—utter *what are you talking about?* shock. I thought that my life revolved around him. That night, I sat alone in my pathetically unfurnished apartment—me, my mattress on the floor, my lamp purchased at a local Target, the spotty wireless internet I jacked from my neighbors, and the half bottle of mediocre white wine in my fridge. I started to call around looking for shoulders to cry on. The response was pretty much unanimous. "I know it's sad, but is this *entirely* a surprise? You've seemed pretty checked out of that relationship yourself."

Still, it was a real low point for me. I was in dire straits financially. I'd wobbled about staying in politics or going into law and made a bad decision. I took the LSATs, paid the fees to apply to law schools, and put down a deposit at Northwestern—the only respectable law school within a manageable drive of St. Louis. (I still don't know why I did that, I guess it was an insurance plan.) My bank account was always one purchase away from an overdraft.

Instead of moving "home" to live with my boyfriend, I moved back to my hometown of Bronxville, New York, to live with my parents—the ultimate humiliation for a twenty-six-year-old. If ever there was a time to reexamine my life and life choices, it was then.

Right before New Year's Eve, the deputy press secretary from New Hampshire, who'd sent me to that Wes Clark event in 2004, and hired me to work in Kentucky, reached out over email: "What about va gov race? Working for McAuliffe?" I responded immediately, "Omg. I just threw up in my mouth. But I'd consider it."

Like a lot of Democrats at the time, I was suffering from extreme Clinton fatigue after the 2008 primaries. I also could never shake my negative feelings toward the Clintons after the episode with Claire. It might have been unfair to judge Terry through that lens, but it's easy to see why I did. He was inextricably linked with them and held the public status of Bill Clinton's BFF. In 2008, he had been Hillary's most enthusiastic supporter. In one of the more memorable TV appearances of the cycle, Terry appeared on *Morning Joe* after pulling an all-nighter, wearing a Hawaiian shirt and brandishing a bottle of Bacardi rum. While media types labeled the appearance "bizarre," "loopy," and "unhinged," I found it endearing and entertaining. Too many politicians take themselves too seriously, and Terry clearly wasn't in that camp. He wasn't *boring*. As I started the interview process with his campaign, I did my due diligence and called friends and colleagues who had worked with him. To a person, they told me: "He is amazing to work for, and *very* fun. You guys will *love* each other."

It was also guaranteed to be a blockbuster of a race. Terry was a national political celebrity, and the off-year Virginia and New Jersey gubernatorial races always garner outsized national attention. Since they're the only two statewide elections the year after the presidential, they're seen as bellwethers for the midterms. The competition to work on them is stiff—they're the only game in town in a job market glutted with unemployed presidential campaign staffers, and they're seen as launching pads for bigger careers.

Eleven days after I received the email about interviewing for Terry's campaign, I was offered the job as his press secretary. The timing couldn't have been better. Nearly overnight, I'd gone from being broke, jobless, and unhappily single to having one of the most coveted communications jobs in national politics. It didn't change my relationship status, but two out of three ain't bad.

There were other obvious perks. McAuliffe's headquarters was based in McLean, Virginia—a hop, skip, and a jump from Washington, DC. I'd never worked or spent much time in DC before then, so it seemed exciting. It also gave me another opportunity, one that I didn't realize the true value of in the moment: to spend more time with my dad.

He'd recently left his job in New York as the vice chair of Sidley Austin Brown & Wood to serve as a senior adviser to one of the commissioners at the Securities and Exchange Commission. When I got the job, he offered to let me stay at his apartment in DC. Sure, it would maybe be awkward at times to be a young single woman living with her dad (and, oh yes, it was), but it would save me a ton of rent money, and he spent every weekend in New York with my mom anyway. I was excited at the prospect of getting to spend more one-on-one time with him—something that had been an impossibility growing up, as I was one of four kids.

Still, the challenges Terry faced were clear from the beginning. His political celebrity cut both ways. His high profile ensured a steady stream of local and national media coverage. And he made for great copy compared to the other candidates. As Jeff Schapiro, the dean of the Richmond press corps, put it: "Say this for McAuliffe: He's added excitement to an intraparty contest that was anything but."

It also didn't hurt that he had the biggest Rolodex in national politics. His fundraising skills were the stuff of legend. During his stint as a fundraiser for Jimmy Carter's 1980 presidential reelection campaign, Terry had wrestled an eight-foot-long, 260-pound alligator for a $15,000 contribution from a Seminole Indian tribe . . . and lived to tell the tale. As chairman of the Democratic National Committee, he was credited with not only saving, but also reviving a dying party organ by raising over $578 million across four years.

On the other hand, a lot of Democrats were skeptical of his

coziness with the Clintons and big-money insider types. When the preeminent national liberal blogger at the time heard I'd taken a job with McAuliffe, he—someone I'd once considered a friend—emailed me: "Yeah, you're now the enemy. McAuliffe is a sleaze fuck who almost destroyed the Democratic Party. I hope he pays well." Jesus H. Christ!

There was also the matter of Terry's roots—or lack thereof—in the commonwealth. A Syracuse native, he'd moved to Northern Virginia just sixteen years prior. While he bragged about his job-creating, deal-making skills, none of them involved businesses in Virginia. And for all his involvement in national politics, he'd had no experience on the local level. Democrats had elected nonnative Virginians before, like Mark Warner in 2001, but people like Warner had at least put in some face time with local Democrats. To the insider class, Terry looked like a loud-mouthed, well-connected outsider trying to buy the governor's mansion.

My first day "staffing" Terry was the day of President Obama's 2009 inauguration. I drove with him and his traveling chief of staff, Justin Paschal, down to Richmond, where Terry would be joining Richmond Democrats for a watch party of Obama's swearing in. Prior to that morning, the only interaction I'd had with Terry was a thirty-second conversation at an inaugural ball the night before. My interview process had been with Terry's senior staff: his communications director (my boss, Delaney, who was responsible for setting the communications calendar and strategy, giving me direction on what to say to the press as chief spokeswoman); his senior strategist (Delaney's boss, Mo, who helped guide the messaging, branding, and larger strategy of the campaign); and his campaign manager (Delaney and Mo's boss, Mike, who made the final calls on strategy and spending). The vibe in the car was icy. Terry and Justin seemed skeptical of me. I was young, I'd never worked in Virginia or national politics before, and

I had fire-engine-red hair to boot. In a walking cliché moment, I'd decided to cope with my breakup by spending eight-plus hours at a St. Louis salon to transform my dyed jet-black hair into what I thought would be a tasteful auburn. The end result was a color of red more closely associated with a dancer at a gentleman's club than a serious political operative. Admittedly, it wasn't one of my finer aesthetic choices.

In the coming days and weeks, Terry and Justin would haze me on a nearly daily basis. They tested me with everything from not-so-nice jokes about former bosses—Claire, Clinton enemy number one, was usually the target—to tall tales about how Terry was related to Christa McAuliffe, one of the astronauts who perished in the failed *Challenger* launch. "You work for me. How could you not know that?" Terry asked with disbelief when I confessed this was new information to me. Justin shook his head silently with disappointment. I couldn't believe that in my comprehensive research before starting the job, I'd missed this fact. When I sheepishly mentioned it back in the office to Delaney, a preternaturally calm veteran of Virginia politics, she laughed heartily: "Um, Lis, you get that they were fucking with you, right?" *Goddammit.*

Our relationship turned the corner one afternoon when we were leaving a policy announcement Terry had made in Richmond. He sat silently in the front seat, listening to me inform a veteran Richmond reporter that I'd "shove [his] balls down his throat" if he burned me on an off-the-record tip. "Lizzy! That's what I'm talking about!" Terry exclaimed once I'd hung up. "GO FOR THE JUG-ULAR!"

It's difficult to quantify all the lessons I learned on that campaign. But the most important takeaway was that the serious work of politics doesn't preclude having *fun*. Every day of my five months with Terry was an adventure. Terry himself had worked behind the scenes of campaigns before becoming a candidate, so he knew the toll that they could take on staff. He did his best to make sure that—outside of the sober policy announcements and stressful moments of crisis—people were having a good time: "Let's light it up!" he'd yell whenever he walked into the campaign office. After even the longest days on the campaign trail, I'd call friends and family to regale them with stories: "Can you believe I'm getting paid to do this shit?"

A month into the campaign, we hired a communications staffer to travel with him and allow me to focus on my role as chief operator of the press. It held for a couple of days, after which Terry called the campaign poo-bahs and told them in no uncertain terms: "I need Lizzy on the road with me."

I could barely contain my excitement. I knew that I was hired to be his campaign press secretary—a job that required my 24/7 focus on speaking to the press—but I loved traveling with him and Justin. The days that I hadn't been with them, I felt an overwhelming sense of FOMO (fear of missing out). I craved the action of the trail, even if it required me doing the double duty of spinning reporters and traveling with and prepping Terry for public events.

The extra work was without question offset by the experiences on the road. We devised a creative plan to generate media coverage: "workdays," when Terry would spend a morning or afternoon on the job with Virginians and allow the media to cover it. We didn't choose just *any* workplaces; we put Terry to work at an algae farm, a chicken-waste-to-energy facility, and firehouses— including one where the firemen were called to put out a grass fire. (In his typical understated way, Terry later claimed that his work that day "probably saved Northern Virginia.") It was catnip for TV and print media: the events fused Terry's natural enthusiasm and extroversion with his campaign message of "jobs, jobs, jobs." One of the recipes for getting good press coverage is making events fun to cover. Reporters are more likely to show up; it's more likely to make the paper or evening news; and voters are more likely to respond positively.

Behind the scenes, Terry brought his prankish side to everyday life on the trail. In the car, there was a strict rule: if you fell asleep, he would take a photo and share it with the campaign staff. "Sleep when you're dead," after all, was one of Terry's life mottos. I made it through all but the last week without dozing off, when my exhaustion finally overtook my pride. Terry and Justin placed a handwritten sign on my corpse-like body and made sure to memorialize it with a blurry BlackBerry photo.

Bouncing around Virginia precluded me from having a traditional social life, but it didn't mean I didn't have *any* social life. Justin and I developed a nighttime routine to keep us from losing our minds. We'd wait for Terry to turn in, then go out—no matter how big or small the town—to patronize the local watering holes in places like Richmond, Virginia Beach, Blacksburg, Danville, wherever. There were a lot of different flavors to Virginia, and we got to see them all. Some nights, we'd be at nice-ish college pizza pubs; others, you could find us at mobile homes that had been converted into gentlemen's clubs, doing our best to stimulate the local economy, one $1 bill at a time. It broke up the monotony of our day-to-day schedules and gave us something to look forward to every night—a new adventure in a new town.

Terry and Justin were both married, so I was the source for all the gossip in the car. I was starting to ease back into single life after my breakup with Jeff. They'd discuss guys they could set me up with and walk through the pros and cons of each: too short, too old, rich but weird. You get the drill. Terry took the assignment especially seriously.

At a northern Virginia fundraiser with hundreds of Pakistani-Virginian donors, he started his speech with a mischievous smile— usually the precursor to a classic Terryism. "I'd like to announce that when I'm governor, Kashmir is YOURS!!!" The crowd erupted in cheers and knowing laughter. No one was under any illusion that the next governor of Virginia would have sway over Kashmir. I don't think there was any doubt in the crowd that Terry would've made the same "promise" to a room filled with Indian-Virginian donors—in fact he had. Terry haters would probably hold up a moment like this as an example of his being a ruthless political operator, but they'd be missing the point. Terry had a guilelessness to him, a quality that made everyone around him feel like they were in on the most inside of jokes.

He scanned the crowd, squinting. When his eyes met mine, I knew I was toast. Though he was from central New York, he'd naturally picked up a hint of a southern drawl after sixteen years of living in Virginia. "Are there any single men in this crowd? Because *mah* press secretary Lizzy over there is just *dah-ing—DAH-ING*—to get married." Every head in the crowd turned in my direction. As a staffer, it's usually a source of pride to get a callout from your candidate in front of a big crowd. That is, unless he's roasting your dating life.

Virginia is massive—it would take nearly nine hours to drive from one corner of the state to the next—so occasionally, we took private planes to attend events. On one flight, we hit turbulence that recalled the legendary scene in *Almost Famous*, in which it seems like the members of the fictional band Stillwater are about to meet Buddy Holly's fate. As they're clearing their consciences by sharing their deepest, darkest secrets, one of the band members cuts the tension by yelling, "We're gonna crash in Elvis's hometown! We can't even die in an original city!" Cue laughter.

It was a similar scene on Terry's plane. While the cabin shook and everyone exchanged wary looks, Terry announced theatrically: "WE'RE GOING DOWN!" He added a visual, his hand plunging downward toward the ground.

"WE'RE GOING DOWN!" he repeated. "YOU KNOW WHAT THE HEADLINES ARE GONNA READ TOMORROW, DON'T YOU? 'MCAULIFFE, OTHERS DIE IN PLANE CRASH!'"

Say what you want about Terry, but he knew how to read a room and diffuse the tension. Everyone—even the most nervous fliers on board—erupted into laughter. Clearly, we all lived to tell the tale.

The final weeks of the McAuliffe campaign were as dramatic as they were brutal. Fueled by his name recognition, aggressive advertising, and wall-to-wall press coverage, Terry leaped to head of the pack early in the race. But his charmed rise finally hit a wall even he couldn't scale.

One of his two opponents—state representative Brian Moran—had decided to stake his campaign's fate on attacking Terry. Moran savaged Terry for his Clinton ties, dubbing him the "booking agent of the Lincoln Bedroom." His campaign ran an ad that claimed, "The truth is, Terry McAuliffe led the campaign that ran the three A.M. attack ad against Barack Obama. The fact is if Terry McAuliffe had his way, Barack Obama wouldn't be our president today." It continued, "McAuliffe even went on national TV and joked, Barack Obama could, quote, 'Kiss my [BLEEP].'" Without the full context of the lively back and forth between Terry and then–*Daily Show* host Jon Stewart, the quote came off a lot worse than it had in real time—downright offensive toward the first Black president.

In another year, Terry's Clinton ties might've been a boost to his candidacy, but in 2009, they most definitely were not. The Democratic Party in 2009 was Barack Obama's party, and a lot of Democrats, especially Black Democrats, were not ready to make nice with the Clintons after the grueling 2008 primary race.

On the other end of the spectrum, the more conservative state senator Creigh Deeds attacked Terry for a mention—just one mention—that he'd made of NAFTA in the epilogue of his memoir, *What a Party*, where he'd listed hundreds of Clinton accomplishments. It was less than an afterthought to Terry, but if you listened to Deeds, you'd have thought that McAuliffe—a Clinton bestie but not an admin official—was an architect of the Clinton administration's key policies. (Note to self: everything you write in a book can and will be used against you.)

Three weeks before the primary, the *Washington Post*—Virginia's paper of record—endorsed Deeds. While Moran and Terry were engaged in a dogfight, the *Post* endorsement was enough to rocket Deeds to the top of the pack. He walked away with the nomination, earning 49 percent of the vote to Terry's 26 percent.

The Clinton haters and McAuliffe detractors were over the moon. A story in the *Los Angeles Times* summed it up concisely: "To many, some with more glee than others, the McAuliffe loss is the last nail in the coffin to the Clinton machine that once catapulted a little-known governor from Arkansas and his Ivy League–educated, ambitious wife into the White House."

I think Terry got a raw deal in that race.

On a campaign level, we made mistakes that reinforced the image of Terry as a big-moneyed, flashy outsider trying to buy the governor's mansion. Terry could raise money like no one else, and we spared no expense. A great example was the annual Shad Planking event—a political confab where insiders dine on overcooked, bony fish; candidates deliver stump speeches roasting their opponents; and the media tries to read the political tea leaves.

It also has some ignominious ties to Virginia's racist past. Coming from New York, I was taken aback by the large number of

Confederate flags at the event. Terry was as well: it was his first Shad Planking, and he was disgusted by the openly racist display. As we departed the event, he vowed: "I'm not coming back to this as governor." (Four years later, when Terry was the presumptive Democratic nominee, he kept his word and skipped the event.)

What stood out most, however, about the 2009 Shad Planking wasn't the Confederate flags—apparently Virginia politicos had become accustomed to the sight—but the ostentatious display from our campaign. In the hours leading up to the event, our field staffers erected more than twenty-five thousand yard signs outside the event, blanketing the final twenty miles of the drive to the woods where it was hosted. We arranged for a plane to fly overhead with a banner reading, NEW ENERGY. NEW JOBS. VOTE TERRY. The crowd was filled with a loud contingent of McAuliffe supporters all sporting blue MCAULIFFE FOR GOVERNOR T-shirts. It looked like astroturfing—political speak for creating the impression of grassroots support.

Our hope had been to show off the organizational strength of Terry's campaign, but we'd overdone it. As we pulled up to the gathering where Moran's campaign was trolling Terry by blasting the Beatles song "Can't Buy Me Love," Terry's wife, Dorothy, winced: "Guys, was this *really* necessary?" Afterward, a liberal blogger wrote up our Shad Planking spectacle as a "vulgar display of wealth." It was hard to argue with the characterization.

It was a classic unforced error on our part, one you see a lot of campaigns make. We were conscious of the negative narrative out there about Terry, and tried to turn it on its head with Virginia Democrats—*Yes, Terry is a prolific fundraiser, but see? It's money that he can put into building up the party and defeating the Republicans in November.*

We missed the mark, and badly. What we saw as a demonstration of organizational strength came across as profligate and tone deaf. There were more effective ways we could have showcased Terry's commitment to using resources to help Virginia Democrats: We could have invested in advertisements in the commonwealth's many struggling small newspapers; we could have hired more organizers on the ground to help Democrats; we could have infused much-needed cash into underfunded county parties—literally anything other than what we did. Rather than dispelling the perception of Terry as an outsider trying to buy his way to Richmond, we reinforced it.

We made other mistakes. When the *Washington Post* endorsed Deeds, they wrote of Terry: "Mr. McAuliffe would be an unpredictable choice, a self-described 'huckster' who has vacuumed millions from donors as a Clinton confidante and former head of the Democratic National Committee."

Terry was apoplectic about the "huckster" line. He'd never used it to describe himself. In *What a Party*, the word he'd used was "hustler"—and he was adamant that we get a correction from the paper. He was, understandably, frustrated with the barrage he'd been facing in the media and wanted a small victory. A few days later, the *Washington Post* tacked a cringeworthy correction to the bottom of their editorial:

> *A May 22 editorial on Virginia's Democratic gubernatorial primary incorrectly stated that Terry R. McAuliffe had described himself as a "huckster." In his autobiography, Mr. McAuliffe described himself as a "hustler."*

Naturally, the media and our opponents had a field day. *Politico's* Ben Smith ran with it on his well-trafficked blog: "CORRECTION OF THE DAY: HUSTLER," noting drily that it was "hard to beat."

U.S. News picked it up as well, delving into the not-so-different definitions of the words:

According to Collins Essential English Dictionary, *a huckster is "a person who uses aggressive methods of selling." A hustler, on the other hand, is defined as "a person who tries to make
money or gain an advantage from every situation, often by im-
moral or dishonest means." Kudos to the* Post *for setting the
record straight.*

Rather than helping Terry, the correction brought more negative attention to one of his biggest vulnerabilities with Virginia
voters. We had committed a cardinal PR sin and unleashed the
Streisand effect, a term used to describe when efforts to suppress
negative information backfire and spotlight it instead. The Streisand effect earned its name in 2003, when Barbra Streisand sued
an unknown photographer for posting aerial photos of her palatial
Malibu estate, citing privacy concerns. Prior to the lawsuit, the
photos had been viewed just a handful of times. Afterward? Hundreds of thousands. News outlets and regular people who never
would have been aware of the existence of said photos were suddenly seeking them out.

The end result was the same in our case. Media reporters and
national outlets ran wild with one throwaway word that we'd insisted be corrected. People who hadn't read the editorial and its er-
rant description of Terry as a "huckster" were suddenly made aware
of it. And the fact that we'd put in the effort to get it corrected
to "hustler"? Hardly a huge upgrade. It made us look foolish and
thin-skinned. It amplified Terry's biggest "negative." Even worse?
We'd driven more eyeballs to an editorial endorsing his opponent.
In the world of public relations, it's tempting to correct every false-
hood and error, but sometimes—painful as it might seem—it's
better to just let it go.

Our campaign may have let Terry down at times, but the media's coverage of him was largely lacking in any nuance. He is,

admittedly, a "work hard, play hard" type of guy. But reading profiles of Terry in 2009, you'd hear only about the "play hard" aspect. Journalists conflated his wild, outlandish persona in his day-to-day interactions with how he'd actually govern, painting the picture of an unserious figure—"Carnival barker" was the favored descriptor. They didn't take a step back and recognize that Terry's energy and his relentless optimism could actually make for very appealing qualities in a chief executive. If he had been the caricature his opponents and people in the media thought he was, he would have taken his ball and gone home after the election. He did the opposite.

A couple weeks after he lost, he got up and started actively campaigning for Creigh Deeds. He put his national contacts to use for Virginia Democrats. He traveled the state relentlessly. Over the next few years, you were much more likely to find him at a gathering of thirty Democrats in Wise County than you were to find him on *Morning Joe*. He put in the time that Virginia Democrats expected of a nominee and, soon after Obama declared victory in 2012, Terry announced his candidacy for governor. He ran largely unopposed for the nomination; he had put in the work, after all. In November 2013, he won the general election with 47.8 percent of the vote to the Republican candidate's 45.2 percent.

Just as important as the fact that Terry won was how he did it. He never stopped being "Terry." He focused himself much more locally, of course, but he was still always the life of the party wherever he went. He was loud and colorful, annoyingly forceful, and always on message. But he always had so, so much fun. And fun is infectious.

Even Virginia Republicans, who were frequently at odds with him, gave him credit. The state Senate majority leader praised his efforts to work across the aisle on his signature issue of job creation: "I think he works tirelessly to promote the state and bring in new business. I'd give him an A." Emmett Hanger, a thirty-year-plus veteran of the Virginia Assembly, admitted: "He's fun to work with."

During Terry's final year in office, he was thrust into the national spotlight when white nationalists of all ilks stormed Charlottesville for a Unite the Right rally where throngs of youngish white men with fashy haircuts carried torches and chanted in unison, "Jews will not replace us." On the second day of the rally, after Terry had called on it to be disbanded, a white nationalist ran over and killed a peaceful Black Lives Matter protester. Visuals from the event dominated the national news and provided Terry with the most high-profile and fraught incident he'd confront as governor.

It was a defining moment of his tenure. The split screen of President Trump defending the white nationalists—"You also had people that were very fine people, on both sides"—and Terry's unequivocal response—"I have a message to all the white supremacists and the Nazis who came into Charlottesville today. Our message is plain and simple: Go home. . . . There is no place for you here, there is no place for you in America"—offered a stark contrast. It was particularly noteworthy as the Virginia media would sometimes paint Terry and Trump as birds of a feather. The *Washington Post* underscored the distinction in a headline days later: "McAuliffe emerges from Charlottesville crisis as a counterbalance to Trump."

In the aftermath of the Unite the Right rally, Terry became one of the first national Democrats to advocate for removing Confederate statues and monuments on public land. It was a radical position to take in a state that was home to the most Confederate monuments in the country. It also put him at odds with the majority of Virginians and Americans on the issue. Seeing the polling on the issue, other Democrats steered clear of it.

Three years later, as protesters took to the streets in anger at

George Floyd's death at the hands of Minneapolis police officers, city after city saw these monuments torn down. Democratic mayors, governors, and presidential candidates raced to denounce the continued presence of these monuments in their communities. Terry may have been ahead of the curve, but finally his position became the de facto one. It was an issue where he showed political courage and prescient moral clarity.

It goes to show you how wrong the conventional wisdom can be. The progressive purists in the Democratic Party had predicted that Terry would be a DINO (Democrat in Name Only) as governor. As it turned out, he was one of the most effective progressive governors anywhere in the country during that decade. There is a destructive tendency in heated, nasty intraparty primaries—see: the 2009 race—to vilify anyone who doesn't agree with you on 100 percent of the issues and label them "the enemy." I'd ask anyone with that mentality to consider Terry's success and rethink their position.

And then, of course, there was the issue of Terry's demeanor—that he was too wild, too brash, and too unserious to be governor. He does shots on live TV! Again, McAuliffe proved his haters wrong. His big personality never took away from the substance of what he was trying to do—if anything, it helped him get more attention for dry policy positions. Also, as he put it in a 2017 BuzzFeed profile: "You want to move people with you? You've got to have fun doing it. People want to be with winners. They don't want to be with whiners. Too many lemon-suckers in this business!"

Terry was the person who taught me the value of joy in politics. I hope his experience also provides a teachable example for Democrats. After Trump's election in 2016, there was a tendency among some in the party to view fun as apostasy, leading us to come across as school-marm scolds next to Trump's freewheeling bravado. I personally think that it's possible to do two things at once: inject a little cheer into our humorless politics *and* fight like hell against the opposition's agenda.

On his last night as governor, Terry threw a rager for the ages. The line outside the door was longer than any I've seen outside the most exclusive Manhattan clubs. Inside, the mansion was packed like sardines with hundreds of Terry's friends, supporters, and former aides who had traveled from across the country and the globe to fete him. Even the New York Knicks legend Charles Oakley was in attendance, towering over the room of political insiders.

"Lizzy, can you believe it?" Terry yelled over the roar of the crowd. His voice was starting to crack from the endless interviews and conversations he'd had in recent days. "This is the house that Thomas Jefferson BUILT! You think he threw any parties like THIS?"

My enduring memory of that night will forever be when everyone was shepherded onto the dance floor for Terry's final speech. He recapped his accomplishments as governor and said:

"Think about this for a minute. It has been a great four years. Two songs to end the governor's reign here in the commonwealth of Virginia. 'Return of the Mack' and 'Hava Nagila.' So I want everybody on the dance floor, you will never see this again in the HISTORY of this mansion. So let's go BIG! EVERYBODY on the dance floor. And let's KICK SOME ASS!"

Terry and Dorothy led the charge in the tent as the opening chords of "Return of the Mack" blared. The rest of us, buzzed from the open bar and high on Terry's improbable route from also-ran to one of the most transformative governors in Virginia history, soon joined in. It wasn't a pretty scene on the dance floor, but goddamn it was a joyful, glorious sendoff.

"Return of the Mack," indeed. And what a satisfying return it had been.

Four years later, Terry would learn a cruel lesson of politics. After he'd served the one term that Virginia constitutional law permitted for a chief executive, he took a four-year hiatus. He returned to run for governor again in 2021. Why not? He'd been popular and gotten things done. Virginia had gone for Biden in 2020 by over ten points. Terry had broken the political curse in Virginia—that the Commonwealth always voted against the president's party in governor's races. Surely, he could again.

Despite a spirited, well-funded campaign in 2021, he lost to his Republican opponent—a first-time politician—by a mere sixty thousand votes out of the nearly three and a half million cast. On any given Tuesday, anything is possible—including a popular former governor losing to a political unknown.

FOUR

The Banker

After Terry's 2009 race ended, I cast around trying to figure out my next job, which didn't take long, thankfully. For good or for bad, politics is a very connections-oriented industry. It's a tiny, impermeable world, where who you know and, conversely, who you've pissed off carries a lot of weight.

The race had been very high profile, and so it wasn't entirely surprising when I started to get feelers from other campaigns in the wake of Terry's loss. First up was the Deeds team, who inquired about my interest in staying in Virginia with a position either on his campaign or with the state party. It might seem awkward for a campaign to try to immediately scoop up an opponent's staffer, but there hadn't been much vitriol between Deeds and McAuliffe.

I also heard from Jon Corzine's campaign in New Jersey. His ad maker had reached out to a friend and asked them to make an intro. It was the other big governor's race that year, and it was already turning ugly by June.

The Deeds campaign ultimately offered me the role of state party press secretary. It was tempting because it would allow me to stay in Virginia, sparing me yet another move and allowing me to work with a press corps I'd gotten to know very well. But prestige-wise, working for a state party is a big step down from working for an actual campaign. It's like the difference between playing for the New York Yankees and the Staten Island Yankees.

I focused my attention on a full-court press for a job with Corzine's campaign. I certainly had some concerns about the race—namely the polling. Even in Jersey, a state renowned for its political messiness, this race looked messy with a capital "m."

Patrick Murray, a pollster with Monmouth University, put it succinctly when he released an April poll showing Corzine trailing Christie 35 percent to 39 percent: "For a Democratic incumbent in a blue state like New Jersey, Jon Corzine is certainly not in an enviable position." By June, Corzine's position was even worse: he was consistently down by double digits in the polls. It was an uphill climb of an election, but I was drawn to the idea of working in Jersey politics, which has a reputation for being as brass knuckles as they come. I knew it would be fun and probably a good learning experience.

I was summoned to Corzine's apartment on a Saturday afternoon in late June for what one of his staffers had described breezily as a "chemistry" interview. I decided to do my homework anyway and read every story about the race that I could get my hands on. The night before the interview, I was up until four A.M. poring over Corzine's record, as well as Christie's public statements and record as US attorney. I took fastidious notes on everything and internalized them.

Thank God, because our meeting that day felt more like an enhanced interrogation than the "chemistry interview" I'd been promised. From the moment two of his staffers met me outside his apartment on the Hoboken waterfront, I realized I was in for a grilling. Why else would the campaign manager and communications director be there in person?

In the elevator up to Corzine's apartment, I was running through the clips I'd read in my head. This was the first time I was interviewing directly with a candidate of this caliber, and I didn't want to blow it. I'd gotten my jobs with people like Claire and Terry by talking with staff, never directly with the candidate.

Corzine answered the door, dressed casually in jeans and slip-on loafers. He cut an intimidating figure, taller and more lithe than I'd anticipated. As I shook his hand, I noticed his icy-blue eyes—icy blue not in the dreamy Paul Newman way, but in the "half dog, half wolf who fatally mauls a child at a roadside petting zoo" way.

Immediately, I was struck by how different he was from Terry—quiet, brusque, and intense. After he grabbed me a Diet Coke from his fridge, he dispatched immediately with small talk, locking his eyes on mine with visible skepticism. He was, after all, the current governor of New Jersey, a former two-term US senator, and the former CEO of Goldman Sachs. I, on the other hand, was a twenty-six-year-old, no-name political staffer with less-than-zero experience in his state. What proceeded was the most rigorous interview of my life.

"So, you worked for Terry McAuliffe?"

It seemed like a normal-enough icebreaker. "Yes, Terry was great, so much fun to work for—"

He cut me off almost immediately: "If he was so great, why'd he lose by so much in that primary? What happened?"

Oh Lord. I did my best to appear unintimidated.

"How much time have you *spent* in New Jersey?" Very little, I informed him. When I mentioned that my dad and sister had gone to Princeton, he shot back in his signature growl: "That's not New Jersey." Point to Corzine there, I couldn't have given a dumber answer.

"Have you *ever* worked with the New Jersey media?" No.

"Do you know *anyone* in the New Jersey media?" Nope.

"So we only have a few months in this race until November—how do you expect me to think that you'll get to know them between now and then?"

I've always had a Jekyll and Hyde aspect to my personality. I could be shy and wary of talking to strangers. But I also could be brash and cocky enough to confront a presidential candidate in front of hundreds of people. Being challenged brought out my competitive side. I knew I didn't know any of the New Jersey media, but I *knew* I could get to. I told him that the same question could have been asked of me with the Missouri, Kentucky, Illinois, or Virginia media, but somehow I'd found a way to make do. Call anyone I worked for, I told him. Finally, he cracked a smile.

"Why do you want to work on this race? Haven't you seen the polls? I'm losing by a lot—don't they scare you off?" I told him that I loved a good fight, and that I'd love to "beat that fucking hypocrite, Chris Christie." I've always allowed myself—a prolific potty mouth—one strategic cuss word per interview. He smiled again.

The rapid-fire questions continued for forty-five minutes. He grilled me on the big issues in New Jersey and Christie's record as US attorney. Thank God for the late-night research expedition. The interrogation got a little bit more comfortable.

But still: this wasn't your usual campaign interview. It was clear that you could take the guy out of Goldman Sachs, but you couldn't take the Goldman Sachs out of the guy. When he'd decided he'd heard enough, he stood up, ended the interview, and ushered me and his two staffers—they'd sat in on the interview—out with little more than a "Thanks, we'll be in touch."

Waiting for the elevator, I turned to the staffers with my heart pounding. "What was that? That was anything *but* a chemistry interview." One of them laughed and in a hushed voice said, "I know it might be hard to tell, but that went *very* well."

It wasn't long before I got the call: I'd gotten the job. When I asked when they wanted me to start, they told me "yesterday." In just eleven days, President Obama would be making his first trip to the state, and they needed everything to go perfectly.

Immediately, I saw that I wasn't in Virginia anymore. Corzine's approach to the press was the opposite of Terry's. Terry loved chopping it up with reporters, staying at press conferences till the bitter end. He'd call over the dedicated Republican tracker, a nice, but clueless college student, and tell him he'd even answer his questions. He invited skeptical liberal bloggers into the car and gave them unvarnished access. Of course, he'd bristle every now and then at an unfair question, but Terry was in his element when he was the center of attention.

Corzine could barely hide his disdain for the New Jersey press corps. He was visibly uncomfortable in their presence and had zero desire to engage with them. He couldn't answer a shouted-out question at a press conference to save his life. It was like he was deathly allergic to giving a coherent and concise response—forget a sound bite. I was told to limit questions at press conferences to two or three. Imagine going from anything-goes press events with Terry, where he would stay until every question had been exhausted, to yelling "last question" as soon as the first one was asked. None of this is meant to diminish Corzine; here's a guy who went from growing up on a farm in downstate Illinois to becoming—through grit, ambition, and sheer intelligence—the CEO of Goldman Sachs. But he was a living, breathing embodiment of a hard truth I was learning about political skills: they weren't things that you could just pick up by reading a book or applying yourself really, really hard. Some people are natural political animals; other people simply aren't built that way.

When Obama made his aforementioned visit to New Jersey to stump for Corzine, I watched thousands of people at the outdoor amphitheater in Holmdel scream at the top of their lungs the second the president—whose approval rating among Democrats was 92 percent(!) at the time—walked onstage. After he concluded his remarks, Corzine returned to the stage and gave Obama one of the most awkward hugs I've seen in my life, seeming to grab on to PO-TUS for dear life. It was a walking metaphor if I'd ever seen one. Corzine was getting brutalized in the polls. His path to victory would rely on clinging to Obama in the blue state of New Jersey. But did he have to do it so *literally*?

My second week on the job, I emerged from the shower one morning to an ominous email from my boss: "Do not respond to any emails or texts. Do not pick up any phone calls. Come into the office immediately." Not the type of email you want to receive on a political campaign—or in any other line of work, or under any conditions, really.

When I got to the office, I learned the cause for concern. A few hours before, federal agents had conducted early-morning raids in the largest anti-corruption sting in New Jersey history. They'd arrested forty-four people, including twenty-nine political officials and operatives—almost all Democratic. You couldn't have scripted a better made-for-TV sting than this one. They cuffed and charged the thirty-two-year-old mayor of Hoboken three weeks—yes, three *weeks*—after his swearing in to office. His crime? Accepting an envelope filled with $5,000 in bribes in broad daylight at a popular local diner. He pled guilty to the charge and apologized to his constituents of twenty-two days about the "disappointment this case has caused." Then there were the five rabbis picked up in the first federal case of organ trafficking—one of whom described himself as the "Robin Hood of kidneys." You gotta admire the moxie—it would be like if Hannibal Lecter referred to himself as the Harriet Tubman of livers. They'd devised a get-rich scheme whereby they'd convince cash-strapped Israelis to sell them their kidneys for $10,000; then they'd flip them on the US market for $160,000. It obviously became immediate late-night TV fodder. It was *so* tawdry. It was *so* New Jersey.

Some of the raids in the case were legitimate; others were not. TV cameras conveniently caught federal agents carrying boxes out of the home of Joe Doria, Corzine's commissioner of the New Jersey Department of Community Affairs. He wasn't arrested, but the visual of the raid convicted him in the court of public opinion. He resigned from his cabinet position that day, because he'd become a liability. The facts didn't get in the way of the Christie campaign's advertisements, which featured the footage of the Doria raid prominently as the "smoking gun" connecting Corzine to the corruption probe. It was pure and utter bullshit. It ruined Doria's professional reputation. Two years after the raid, the new US attorney took the rare step of sending a letter to Doria's lawyers letting them know that investigators would not be filing any charges against him.

There was another thing that stunk to high heaven. The day of the arrests and charges, Christie was set to hold a public event in Hudson County. At the Corzine headquarters, we'd scratched our heads the evening before when we saw his public schedule—*Why on earth would Christie go to Hudson County?* Hudson County is one of the most reliably Democratic areas in New Jersey; campaigning there as a Republican was a complete waste of time. It would be like Joe Biden stumping in rural Oklahoma, or Donald Trump holding a rally in the Castro in San Francisco.

Once the raids went down, it all started to make sense. The political arrests were concentrated in Hudson County. The timing of Christie's public event couldn't be a coincidence; our assumption was that he'd been tipped off. He milked the shit out of the moment, giving an Academy Award–winning performance for the wall-to-wall press that showed up at his West New York presser:

> *This is obviously just another really tragic day for the people of New Jersey. Over the last seven years, I worked extraordinarily hard along with the career professionals at the U.S. Attorney's office at the FBI to combat the culture of corruption in this state, and unfortunately today is another example that there is much work to be done.*
>
> *It's a bad day for the citizens when they are once again disappointed by their public officials. But the fact of the matter is we should all feel gratified about the fact that we have career prosecutors at the U.S. Attorney's Office, and career professionals at the FBI and the IRS who remain vigilant about doing their job, and doing it the best and most professional way they can.*

It was as brutal a workday as I can remember. The only comic relief came when Justin forwarded the news alert from the *New York Times* to Terry and me: "What kind of state are you running? Are the mayors and other pols dems? If so, you're fahkkkkkked." Terry responded with a succinct: "Wouldn't happen in my VA!" God, I missed them.

Justin's assessment of the race pretty much reflected the conventional wisdom: Corzine was *fucked*. A week later, Stuart Rothenberg wrote a column titled: "Four Months to Learn to Say 'Governor Chris Christie,'" where he described the raids as "the straw that breaks [Corzine's] back in November." His conclusion was unequivocal: "When I asked one longtime Democratic insider about the race, it took him all of two words to assess Corzine's prospects: 'It's over.'"

On the ground and in the campaign office, it didn't feel "over," but close to it. The rumor mill about Corzine dropping out to clear the way for a better candidate reached a fever pitch. The White House was making the calls to other New Jersey politicians, like Cory Booker, seeing if they'd potentially step up as a replacement.

Corzine stayed in the race, but for weeks, it felt like a death march. Albeit a very, very wild one. Jersey politics lived up to the hype: they were different from anything I'd experienced before.

That summer, as Corzine was set to make an appearance at a gubernatorial forum, there was a very loud contingent of Republican protesters hurling epithets and chanting against him. One of our staffers called over the local labor leader present and whispered in his ears. Within minutes, half a dozen burly labor guys with bagpipes drowned out the noise of the protestors, allowing Corzine to enter the site with a little more dignity. I still have questions about it to this day. At the top of the list: do labor guys in New Jersey just carry around bagpipes in their trunks like they're spare tires?

At the time that Corzine was battling the polls, I was dealing with a crisis of my own, one that I did my best to hide from my co-workers. After Jeff and I had broken up, we'd basically stopped talking; it had been almost nine months since we'd spoken. So I was surprised in late July, just after the anti-corruption raids, when I heard my BlackBerry vibrate with a call from him.

He intimated that some big news was going to break, but he wouldn't share any details with me—it was just a "heads-up" call. The usually confident Jeff sounded distressed and shaken. Over the next few days, I bombarded him with texts: "What's wrong? I know something's going on." Finally he came clean. He'd been caught up in an anti-corruption sting of his own, also at the hands of a Republican US attorney. Not only that, he was about to be indicted—and he was going to resign from the Missouri Senate.

Some of the details Jeff shared weren't *completely* new to me. When we first got together, Jeff had mentioned a Federal Election Committee (FEC) complaint that his 2004 opponent, Russ Carnahan, had filed during that election. The complaint alleged that Jeff and his campaign were behind anonymous mailers attacking Carnahan's attendance record as a state representative. The mailers were in violation of FEC rules because they didn't identify who paid for them, as required by law. Given that they hit twenty-five thousand voters' mailboxes at the time that Jeff was making the same attack in the press, the Carnahans suspected his campaign was behind it. And they were right. Still, in a sworn affidavit to the FEC complaint, Jeff denied any knowledge of it and the FEC dropped the case. It was a stupid mistake, borne out of a mixture of panic, inexperience, and plain bad judgment. Nowadays, most candidates and campaigns facing FEC complaints of their own have found a workaround: they just waive their right to a defense. The FEC is so toothless that a non-defense usually leads to no action.

It's an uncomfortable reality that a lot of campaigns play fast and loose with election laws, betting that the FEC will never actually catch up to them. Around the same time in 2004, the Swift Boat Veterans for Truth was smearing John Kerry's Vietnam War service with a book, TV ads, and untold amounts of free media coverage. The group was funded by some of Bush's biggest donors and family friends; it shared a general counsel with the Bush campaign and had countless ties to Bush's "brain," Karl Rove. I don't think anyone in their right mind would look at that effort—

one which likely swayed the outcome of the entire presidential election—and think that it was completely kosher, but it never resulted in anything other than a lone fine from the FEC.

That's why I was so surprised when Jeff told me that he was about to be indicted for lying to the feds about the whole ordeal: the case was closed in my mind. But it turned out that in 2008, the guy who produced the illegal mailer, Skip Ohlsen, was picked up by the FBI for planting a bomb in a St. Louis–area parking lot. He'd intended to kill his ex-wife's divorce lawyer but put the bomb under the wrong car. It didn't take long for the FBI to zero in on him as a suspect and build an airtight case.

Clearly, someone who is involved in—and botches—both an anonymous mailer campaign and a car bombing in the course of four years is not playing with a full deck of cards. But Ohlsen had one card to play with the feds: he could implicate someone more important than him in exchange for a reduced sentence for the bombing gone awry. That someone was Jeff—the biggest rising star in Missouri politics. He was also a Democratic elected official, and there was an insatiable bloodthirst for prosecuting Democrats at the hyper-political, Bush-era Department of Justice.

By the time the FBI knocked on Jeff's door, they'd already been taping his conversations with others caught up in the investigation, and they'd gotten him on tape admitting to knowing about the mailer and lying to the FEC. His goose was cooked.

I was in complete disbelief. A *car bomb*? A *wire*? Jeff, of all people, about to be *indicted*? It was more absurd than the plot of a Coen brothers' movie. Except it was real life. After an hour on the phone with him that night, I sat in the parking lot of my apartment building in Ewing, New Jersey, crying uncontrollably in my car.

Even post-breakup, I still cared for Jeff. I knew how passionate he was about his job. I'd seen the effort he'd put in to get there—

all those long days knocking on doors and late nights writing notes in bed. Most of all, I knew that he was so much better than the 99 percent of the other people who serve in state legislatures—smarter, harder-working, more innovative, more likely to get things done. Yet now, a series of seemingly small, inconsequential decisions was about to end his life's dream of serving in public office.

I wanted to be there for him. We went from not talking at all to suddenly talking all day, every day. I compiled lists of people who could weigh in on his behalf with both the press and the court and drafted talking points for them to hit. I worked with him on what he would say when the indictment became public. And I made some calls to other targets of the investigation that I probably shouldn't have.

Even though he was facing a significant sentence, Jeff refused to play ball with the FBI and the US attorney. He wasn't going to wear a wire to catch a bigger fish; he wasn't going to deflect blame onto others; he wasn't going to be a snitch. No matter what they asked, no was his answer. He was flawed, but he was also a person of principle. He wasn't going to cooperate with the feds and make other people suffer to reduce his prison time. He could deal with the consequences of his own actions.

My stress levels were off the charts. I couldn't tell anyone—family, friends, or work—what was going on. It was made worse by the fact that all day long I was sitting in an office that was gripped by hysteria over the local US attorney's investigation. We wondered if our conversations were being listened to, and went outside the office—just in case—to hold sensitive conversations during smoke breaks. In the back of my head I wondered—if they're this paranoid about everything here, how would they react if they knew I was connected tangentially to another anti-corruption investigation halfway across the country? One that was about to hit the national news?

Around that time, I was driving to work one very slick, rainy morning when my Nissan Xterra skidded on a sharp turn and hit a telephone pole. Both my car and the telephone pole were destroyed, but I was physically unscathed.

I was still in shock when the responding police officer arrived at the scene—I'd been in a couple fender benders in my life but never anything like this.

He ran through the normal procedures. He took a Breathalyzer that showed my BAC at 0.0 percent. *Obviously.*

He asked if I'd been on the phone or texting at the time. I hadn't, and I offered up my two BlackBerrys as evidence. *Good.*

"So what happened, then?" he asked. *Not good.*

I threw up my hands in despair, shook my head, and told him the truth. "I don't know. I don't know. I guess I just lost control." I rented a little red Hyundai sedan from the local Hertz for a couple weeks, and then placed a down payment on the car I still drive today.

In reality, I was losing control of every aspect of my life. It was evident to everyone around me. I was exhausted and distant at work, and I sometimes struggled with the most basic tasks.

My boss cleared the office one day and called me in for a "come to Jesus" talk. He was incredibly sympathetic and tried to connect with me on a personal level. "Lis, you were doing great work at the beginning of this campaign. But you're just not performing up to your potential anymore and not doing what we need you to do. Is something going on?"

Something was. It just wasn't something I was willing to share. I avoided his eyes. "No, nothing's going on with me." It was a brazen lie.

My boss told me kindly but unequivocally: "Either start working at the level we need you to, or we'll find someone who can." Never before—and never since—have I received such a professional reprimand.

It was an invaluable learning experience. What I thought came across as slight detachment translated as total incompetence. My efforts to hide the extreme emotional duress I was under made me seem evasive, at best. And like a secret junkie, at worst.

To this day, as a woman in the workplace, I still struggle with how to appropriately handle moments of personal vulnerability. There's no good answer, no matter what HR tells you. The tough reality is that no one likes a woman who cries or shows weakness.

If I had a time machine, I would go back and ask for a few days off. I badly miscalculated my ability to compartmentalize the pain and stress I was feeling and learned the hard way that admitting vulnerability is better than projecting failure.

Everything in my personal life felt like it was going to shit. Everything in my professional life felt like it was going to shit. I couldn't detect even a glimmer of light at the end of the tunnel.

But then, out of nowhere, our campaign got a gift from heaven. Just one month after the raids, Zack Fink, a statehouse reporter for the New Jersey Network (NJN), alerted our campaign to tune in to the evening broadcast—he had a bombshell story coming within the hour.

At the Corzine headquarters, staffers crowded around a small TV and watched eagerly as Fink reported that, as US attorney, Christie had lent tens of thousands of dollars to a subordinate, Michele Brown, and failed to report the loan on his taxes as required by law. It was significant for a number of reasons.

Obviously, breaking the law is never a good look for someone running for office as a high-and-mighty, squeaky-clean crusader. But it also opened the door to a whole range of other questions about Christie, Brown, and the US attorney's office—questions reporters were *finally* interested in asking. What was the nature of Christie's relationship with Brown? What was the nature of Christie's relationship with the US attorney's office? Did he have any advance notice of the corruption raids? Was the office taking concrete actions to help his campaign?

These were questions that our campaign had been pushing to reporters for weeks and months, and not baselessly. The Corzine campaign had compiled reams and reams of opposition research—political speak for "dirt"—on Christie. It showed that he wasn't exactly the virtuous reformer he wanted people to think he was. Up until now, our pitches and research had mostly met with shrugs and "Who cares?" from the press corps. They'd seen the polling and absorbed the conventional wisdom that this race was "over." Why waste any time paying attention to oppo in a race that was becoming less competitive by the day?

Zack's story blew the whole thing open and shook reporters out of their complacency. Suddenly, everywhere you turned, there were critical stories about Christie. The *New York Times* ran a front-page piece on how Brown had used her role in the US attorney's office to help Christie's campaign. Other outlets picked up on how he'd flashed his badge to get out of traffic violations and gotten verbally abusive with officers who didn't let him off the hook. And then there were all the stories about his time as US attorney—how he'd used the office for political retribution and blown taxpayer dollars on limo rides and rooms at the Four Seasons.

The drip-drip-drip of bad stories slowly took Christie's greatest strength—his perceived anti-corruption bona fides—and turned it on its head. He began to look more like a sleazy pol on the take than a white knight. It was a dream scenario for our campaign. We were never going to win the race by convincing people that

Corzine was the bee's knees. We'd lose a referendum on Corzine and his record, but we could win one on Christie's. The public polling began to shift, and by late September, Corzine had gone from being a dead man walking to being in a dead tie with Christie.

We didn't always get it right. Even in New Jersey politics, sometimes you can go too far. We knew Christie was sensitive about his weight, and our tracker would regularly send videos from the campaign trail capturing Christie's tortured relationship with food. In one, he got Christie staring ruefully at a box of doughnuts he'd been handed at an event. In another, he caught him making a pit stop at a McDonald's. Right as Christie was about to walk out to his State Police detail with bags of fast food under his arms, an eagle-eyed staffer spotted the tracker in the parking lot. Christie handed off the bags to the staffer and walked outside as nonchalantly as possible, sipping on his soda. He was truly that paranoid.

It all culminated with a Corzine ad alleging that Christie "threw his weight around" as US attorney. The line was accompanied by a slow-motion visual of the corpulent Christie exiting the front seat of his SUV. The *New York Times* deemed the ad "as subtle as a playground taunt," noting how our campaign videos often featured "unattractive images of Mr. Christie, sometimes shot from the side or backside, highlighting his heft, jowls and double chin." (In fairness, none of the videos were doctored.)

We put the ad into heavy rotation, and its subliminal message sunk in with New Jersey voters. Shortly after we released it, Monmouth University conducted a poll asking respondents the first thing that came to mind when they thought of Christie. The most popular response? "Fat."

Still, the ad was roundly criticized for "fat shaming" Christie. It elicited sympathy and allowed him a humanizing moment to discuss his weight struggles. He opened up at a campaign stop, telling the supporters and media assembled:

I'm gonna let you in on a little secret. I know most of you didn't know this—but the governor's been whispering this to the press for months and months and months, and now he's trying to be a little cute about talking about it too through his TV ads. I want to make sure you're all seated and you're OK before I let you in on a secret: I'm overweight. And I've struggled with my weight for the better part of 30 years, up and down.

And the governor somehow thinks that in a time when we have 9.8 percent unemployment, at a time when we have the highest tax burden in America, at a time when we have the highest property taxes in the United States, that that's what you wanna talk about.

Point to Christie. Looking back on the episode now, it's hard to justify our focus on Christie's weight. It was insensitive and juvenile. There were plenty of legitimate criticisms of Christie—it's not like we were at a loss for material. Instead, we chose to zero in on a struggle that he shared with a lot of New Jerseyans.

It serves as a cautionary tale about the bunker mentality that can consume a campaign. To us, Christie was a bully. He was the enemy. He would routinely launch over-the-top attacks at Corzine and other political opponents. If he could dehumanize his opponents, why couldn't we? It calls to mind a famous quote frequently attributed to George Bernard Shaw—and no, this isn't me making a fat joke—"Don't wrestle with pigs. You both get filthy and the pig likes it."

Years later, I saw a similar dynamic unfold in the 2016 Republican primary, when Donald Trump launched nasty personal attacks, the likes of which had never been seen in presidential politics. He ridiculed Jeb Bush as "low energy," referred to the shortish Marco Rubio as "lil Marco," accused Ted Cruz's father of being involved in the JFK assassination, and even retweeted a post juxtaposing an unflattering photo of Cruz's wife with a modeling photo of Melania Trump, captioned: "A picture is worth a thousand words." Trump's opponents initially blew off his attacks, but as the GOP contest got down to the wire, they removed the gloves.

In March, Rubio opened a rally by attacking Trump: "I'll admit, he's taller than me. He's like 6'2", which is why I don't understand why his hands are the size of someone who's 5'2". Have you seen his hands? . . . And you know what they say about a man with small hands." Rubio might as well have just come out and accused Trump of having a small dick. Not content to leave it there, Rubio took aim at Trump's peculiar orange skin tone, saying, "He doesn't sweat because his pores are clogged from the spray tan."

Rubio's crass comments did him no favors. Unfair or not, it was accepted that Trump was a boorish bully. People didn't expect better from him. When Rubio stooped to his level, he looked—no pun intended—small and petty. He undermined his own brand as a fresh-faced, different kind of Republican. He wrestled with a pig and came out of it filthy. The same could be said about our approach to Christie with that ad campaign.

We went into Election Day with complete uncertainty as to how it would play out. Of the last two public polls released in the race, one showed Corzine leading by 2 percent, while the other had Christie up by 3 percent. Our own internal polls, conducted by President Obama's pollster, painted a rosier picture, predicting that Corzine would win by 5 percent. In the end, Christie edged out Corzine by a margin of 3.6 percent.

It sucked to lose, especially after how close we'd come. There was righteous anger within our team that Christie had played dirty pool to win, seemingly weaponizing the US attorney's office in the race. Still, the general consensus was that Corzine had over-performed expectations and run a pretty good campaign. In contrast, Deeds had gotten absolutely shellacked in Virginia, losing by 17 percent.

I'd made the decision to work on the New Jersey governor's race over the then-safer pick in Virginia in part because I believed it would be an educational experience. And boy was it ever. At times, maybe we had gone too far, but our willingness to get down in the mud, even if it dirtied us up a little bit, was central to Corzine's comeback. Too often, politicians and their handlers talk themselves into the value of taking the high road and running a "clean" campaign.

There's no such thing as a moral victory in politics. Either you win or you lose. "Moral victory" is usually just a euphemism for losing.

We might have crossed a line in the Corzine campaign going after Christie's weight. But more substantive lines would be crossed in years ahead.

In his first year in office, Christie became a national political phenomenon. It all started with a town hall where he shouted down a New Jersey teacher who dared to challenge his routine denigration of public employees. The clip went viral, and Christie's office realized that they'd stumbled on a magic formula. He held regular town halls across the state, and at each and every one, he'd find at least one target for his ire. Cable news would cut into the town halls live, where political pundits and journalists would fawn shamelessly over them. Within twelve months, Christie was frequently mentioned as a 2012 presidential candidate.

It was politics as performance art, and it was troubling how much of the media put a premium on the clips' entertainment value. Here they were, championing and amplifying Christie's most abhorrent qualities—qualities that we teach children not to embody. He was cruel, he was rude, and he was a total, total asshole. It may have made for great TV, but to what end? It certainly didn't serve any public interest. The focus on Christie's style came at the expense of coverage of the substance of his administration. He killed the ARC tunnel, a major rail project connecting New Jersey to New York City that would have created thousands of new jobs and reduced the notorious congestion that New Jersey commuters contended with daily. He routinely racked up large budget deficits. He cut funding for New Jersey schools and law enforcement to pay for tax cuts for wealthy New Jerseyans.

And yet Christie had carved out a unique niche in American politics, it seemed. Surely no one could be more grotesque or more offensive, right?

In 2016, Christie saw the monster he helped create turn on him. His dime-store Tony Soprano shtick paled in comparison to Donald Trump's hate-a-palooza. Trump was louder, cruder, and even more offensive. Like Christie, he didn't always fall in line with the conservative voters in the Republican Party, and, like Christie, he was able to appeal to the far right by making clear that no one would put more energy into attacking the people they hated. For Christie, the target had been public employees and teachers' unions. Trump found an even better scapegoat: immigrants.

Christie never stood a chance, and he soon found himself on the outside looking in as the media breathlessly covered Donald Trump's every word. The cameras that had previously cut into Christie's town halls were now glued for as long as an hour to the empty podium awaiting Trump at his raucous rallies.

There's no doubt that Christie saw a little bit of himself in Donald Trump. And yet for all the talk of Christie being a takedown artist at the debates, he never once laid a finger on Trump, the undisputed front-runner. Instead, he trained fire on easier targets like "lil Marco." Shortly after he dropped out of the race, he became the first major Republican to endorse Trump.

The media has engaged in some much-needed soul-searching in the wake of the 2016 election. But their practice of puffing up boorish bullies started long before Donald Trump.

FIVE

The Psychologist

At this point, I was coming off four straight losing campaigns. Friends outside of politics worried that these losses would affect my ability to get hired for future jobs. Nothing could have been further from the truth.

In politics, your win-loss record only really matters if you're the candidate or maybe the campaign manager; otherwise, you're judged on the quality of your work and whether you were able to move the needle for your candidate. There's also just a lot more respect for people who take on tough, losing campaigns than for people who work on easy wins. It shows mettle, and the experience you take away from a difficult campaign—dealing with crises, negative stories, nonstop attacks, etc.—is invaluable. If you're a GM of a football team in need of a quarterback, are you gonna draft the amazing quarterback who played for the team with a losing record or the okay one who played for the team with a winning record? It's no different when you're building a campaign team.

The hard part was making a choice about where to head next. I narrowed it down to two races: either the Ohio US Senate race . . . or the Ohio gubernatorial race. I *really, really* wanted to work in Ohio, and it wasn't because I was dying to live in the Buckeye State and go to OSU football games. Times have changed a bit, but in 2010, Ohio was *the* holy grail of American politics. "As goes Ohio, so goes the nation" was a famous political saying for a reason. No Republican had ever been elected president without winning Ohio, and the state had determined the winner of every presidential election since 1960.

There are other reasons to work in Ohio. It's seen as a microcosm of the United States because of its regional and economic diversity. Working in Ohio, you get a taste of big-city, suburban, rural, and Appalachian politics and culture. Outreach to Black voters is as critical as outreach to white voters. And to win the state, you've gotta have a deft political strategy—you can't just rely on turning out your base. You've gotta appeal to the other side as well.

To the casual political observer, Senate races may seem "sexier" than gubernatorials. In most people's minds, there's a lot more cachet to working in Washington than working in a state capitol. But Senate races come with a lot of constraints. For one, you're pretty tethered to what your party is doing in Washington. If Democrats are having a rough go on Capitol Hill, it's likely that you will on the campaign trail. Gubernatorial campaigns are a different beast—you can create your own brand and identity, completely distinct from the national party. The Democratic Governors Association, the gubernatorial counterpart to the DSCC, actively encourages candidates to distance themselves from alienating policies in DC. And in the grand scheme of things—governors are much, much more powerful than US senators. The governor of New York, for instance, oversees a $200 billion annual budget and a 250,000-plus-person workforce. A senator from New York, on the other hand, is just one of a hundred votes in the chamber, and they have—at most—a few dozen staffers in their office.

I went back and forth between the two races. A lead staffer for the DSCC told me, condescendingly: "You can either be a hero in Washington or Columbus, Ohio. It's up to you." I went with my gut—the governor's race it would be.

Corzine might have seemed like a different breed of politician from Terry. But Ted Strickland, the governor of Ohio, was an entirely different species. He'd grown up one of nine children in Appalachian Ohio in a place called Duck Run. A tenth sibling that he rarely talked about had died at a young age. His dad was a steelworker who had a fifth-grade education. When Ted was four years old, his house burned down, and the eleven members of his family had to live in a chicken shack for several months. Despite this adversity, Ted went on to be the first member of his family to get not only a college degree but also a master's in guidance counseling, a master of divinity, and a PhD in psychology. Before he was elected to Congress and as governor, he'd worked as a minister and prison psychologist. It was as unconventional a political bio as I'd ever seen.

The characterizations his ex-staffers provided in advance of our first meeting were as foreign to me as his résumé: "He's probably the kindest person you'll meet in politics" or "He's the best human being you'll ever work for" were typical. And yet they were quick to qualify their assessments with reassurances along the lines of "He is no dummy. He is a *political animal*."

When I started, I had a longer runway with Ted than I did with Terry or Corzine. On those campaigns, I had four or five months to prove myself. They'd have to either love me or revile me pretty quickly. Here, I'd have ten months.

I was introduced to Ted when he dropped by the campaign

office during my first week on the job. The big space we'd rented out a few blocks from the state capitol was largely unfurnished and empty—it was the early days of the race, before the staff had filled out. "The big dog's coming in," Aaron Pickrell, our campaign manager, informed me right before Ted waltzed in surrounded by the three state troopers who were always by his side. Ted and Aaron exchanged some inside jokes, laughing with each other as I sat awkwardly off to the side. Finally, Aaron introduced me: "Ted, I'd like you to meet your new communications director, Lis Smith."

"Oh yes, I've heard about you," Ted responded in his trademark lilting voice. "You've come here from *Washington* to help save *Ohio*." The gentleness of his tone didn't take the sting out of the verbal slap that he'd delivered.

As soon as I started to correct him, I regretted it: "I'm actually from New York . . ."

"Oh, so you've come from *New York* to *Ohio* to save us?" *Wince. Wince. Wince.* Clearly it would take some time for him to warm up to me, but he eventually would.

As billed, he was gentle and kind—if a bit standoffish at first. He and his wife had never had children, and he treated his closest staffers and even his state police detail like members of his family. He never raised his voice or got mad. One day, when his body person—campaign speak for the staffer who spends every waking minute with the principal—didn't show up for a seven A.M. call time before a long day of travel and campaigning, Ted didn't even flinch: "He's been working so hard. He probably deserves some extra sleep." Most candidates would've blown their gasket; some would've even fired the staffer. Instead, Ted just turned to me with a smile: "Looks like you'll have some extra work today." When the staffer called me in a panic, Ted took the phone out of my hands and told him not to sweat it. He'd get by just fine without him that day.

PhDs are rare in electoral politics—at the time, Ted was the only governor in the country who had earned one. Working with Ted up close, I came to understand the value of that psychology degree.

Politics is a confounding business. Dealing with the quirks of all the players—the elected officials, the donors, the lobbyists, the reporters—is one of the most frustrating parts of the job. Candidates—people who had worked in the business for years—would sometimes turn to me and ask incredulously: "Why is [so-and-so] doing this?" That was never a thing with Ted. The psychologist in him was always at work—analyzing the different personalities he had to deal with, assessing their strengths and weaknesses, and sussing out their motivations.

That Democratic politician who had given a less-than-gracious quote about him in the paper? "You just need to understand that this guy is very insecure about his stature. He's just overcompensating for it by being loud and brash."

He could be more blunt at times. That weirdo reporter who wouldn't make eye contact in interviews? In debate prep, Ted gave a roomful of his top advisers his assessment of him: "You know what his issue is? He feels shame. I bet he masturbated a lot as a child."

"No, Ted! Stop! No!" a few of us yelled as we lapsed into hysterics. The visual was too much.

Ted continued, unfazed and now grinning like the Cheshire cat: "Yes. He masturbated a lot as a child and he was *very, VERY* ashamed of it."

There was a new political reporter at the *Toledo Blade*, Joe Vardon, who had written a couple of very negative stories about Ted. Some members of the campaign team were adamant that Ted should not engage with him, that we should freeze him out completely. Ted had a different point of view: "Oh, Joe's adversarial because he's young, hungry, and wants a scoop. He's smart. He's not a bad guy. Let me just talk to him."

A few days later, I was with Ted on a multi-city swing through northeast Ohio when we stopped to get coffee at a Wendy's. The press who were following Ted's detail pulled in after us to enjoy the spectacle of the governor of Ohio in line at a fast-food joint. Joe was one of the reporters; the others were more senior and established members of the press corps.

When we all left the Wendy's to get back in our cars, Joe threw out a seemingly absurd request: "Hey, Ted, why don't you ride with me to the next event?" The ask itself seemed ridiculous on its face—surely the governor, with his two-car state police detail, wouldn't just jump into the front seat of this young reporter's car. The other reporters laughed at his chutzpah—*as if!*

"Yeah, why not?" Ted responded. "Lis, come on—get in with me!"

The other reporters' jaws dropped. *How the hell is this young guy getting access to the governor that I never get?* (One of them later told me that it was hard not to take it personally. I asked him how many times he'd ever asked the governor to get in the car and, of course, the answer was zero.)

Ted's assigned security detail tried to intervene: "Are you sure this is the best idea?" It was *their* job to drive the governor around and keep him safe.

Joe's eyes widened, too. He had been joking, but suddenly he was like the dog that caught the car. *Now what?* Well, first, he had to clean out what looked like weeks' worth of fast-food wrappers and other detritus from the passenger's side of his red Pontiac sedan, so that the governor wasn't sitting in literal trash. "It's a company car," Joe told him sheepishly. It was a great scene—the wounded, jilted reporters, the horrified state troopers, the shocked cub reporter, and Ted standing there with a big ole smirk on his face.

"So, *Joe*. Tell me about *you*," Ted said as Joe turned on the ignition.

"What do you want to know?"

"I want to know about your family. I want to know about *you*. How did you get to where you are now?"

For thirty minutes, as Joe drove us to the next event, I fielded worried texts from Ted's state troopers, who were closely tailing the car. *Yes, Ted is wearing his seat belt*, I told them. *Yes, we can trust Joe not to drive his car into a brick wall on a suicide mission.*

Halfway through the drive, Joe turned the conversation back on Ted: "I want to know about *you*. How did *you* get to where you are today?"

Joe knew all about Ted, but there's a difference between interacting with a politician at a press conference and talking with him in the front seat of your trashed car. It was all on the record, with a couple exceptions. Ted went off the record a few times to use more colorful language than the *Toledo Blade* could run in print.

And you know what was crazy? It was one of the best conversations I'd ever heard Ted have with a reporter. I didn't alert the higher-ups on the campaign until the car ride ended—I knew they'd try to torpedo the idea or question my judgment. When we arrived at our stop and got out of the car, Ted turned to me and asked: "Now tell me, why haven't I been doing that for the last four years?" Inside, Joe pulled me aside to tell me how much he appreciated the time: "I feel like I get the guy now. You know, he's really good in that context. I didn't buy all of his answers to my questions. But you should do that more." It didn't make Joe pull any punches in stories going forward, but it added more context and balance to them. It's easy for a reporter to take an unfair shot at a politician if they just see them as a cardboard cutout. It's harder to sucker punch a real, three-dimensional person.

I wasn't exactly a tyro at dealing with the press when I'd joined Ted's campaign. I'd witnessed a lot in my six years in the trenches, and I was pretty sure I had it all figured out. But to this day I still think about that one seemingly small moment and what it taught me. In Ted, I had a candidate who was uniquely intuitive about human nature. A part of him would always be the prison psychologist who could find the best, most redeemable qualities in people. He could relate to almost anyone. Most politicians don't have that skill set.

Neither do most people in my line of work. In the heat of a political campaign—or really any PR shitstorm, if we're being honest—it's easy for people on my side of the fence to view reporters and the media in general as a monolith. They hit them up with identical pitches; they talk to them all the same way; they yell at them for an unflattering story or even just one unflattering sentence in a story. Rinse and repeat. It's just a horrible, lazy model for communicating, and it's extremely common in the political industry. Ted reminded me to always think not just of *the reporter*, but of *the person* I was talking to. *Who are they? Where are they from? What makes them tick?* We can't attack the media for covering politicians as cartoon characters if we are similarly guilty of reducing reporters to Wile E. Coyote.

In a lot of ways, Ted embodied the opposite of the Democrats I'd worked for. He was an ordained minister, don't forget, and when he'd get into fiery pastor mode, he would get people on their feet, whether it was at a Black church in Cleveland or before a crowd of camouflage and bright orange–wearing outdoorsmen in Appalachia. He was down-the-line anti–free trade, going back to his votes against NAFTA and free trade with China, even though those policies had been pushed through under a Democratic president. And he had an A+ rating from the NRA—something that in 2010 was rare for a Democratic politician, but which would now be inconceivable. We traveled the state in a camouflage RV to tout his record of standing up for hunters and sportsmen. It was about as far from New Jersey as you could get, where Corzine had run attack ads against Christie for "standing with the N.R.A." It was total political whiplash: within a matter of months I was arguing the virtues of positions that months before I'd been attacking.

There are people both in and outside of politics who would have a problem with doing that. *How could you work with people with such different views? Don't you have any core beliefs?* I've heard every iteration of those questions over the years.

The simple answer is that I do have my core beliefs, and while they may have gotten more nuanced over the years, they haven't fundamentally changed much. But they're *my* beliefs. I'm not so arrogant and close-minded that I think everyone needs to share them. If I'd gone into politics to work only for people who shared 100 percent of my policy views, I would've spent the last eighteen years of my life wandering the political wilderness, searching for a unicorn.

But there's something even more important than that—something that Democrats (and Republicans, too) can sometimes forget. A Democrat who can win in a liberal district in New York City looks a helluva lot different from a Democrat who can win in upstate New York, let alone in West Virginia or Ohio. In recent years, especially, there has been this dogma of ideological purity pushed by some on the far-left wing of the party. If someone doesn't support every policy on their progressive wish list, no matter how fanciful or unfeasible, no matter how politically toxic, they're branded an enemy or a Republican in disguise. In my experience, the *only* thing that smug, know-it-all attitude produces is more Republicans in office. And, oh boy, if these ideological purists think a West Virginia Democrat is bad, wait till they get a load of the Republican alternative. In 2011, I'd help elect a self-described "100 percent pro-life" Democratic governor in West Virginia. Despite his "pro-life" views and endorsements, he still vetoed a twenty-week ban on abortion, citing constitutional concerns. In contrast, his Republican successor signed the most extreme antichoice bills that came across his desk, including a "Born Alive" bill, which basically painted abortion doctors as murderers.

The Ohio campaign closed with more drama than the Corzine race. By 2010, the national political landscape had shifted even further to the right. Even if Ted didn't have the personal baggage or political limitations that Corzine had, he wasn't immune to the coming Tea Party wave. A September poll showed Ted trailing his Republican opponent, former congressman John Kasich by 17 percent; he looked dead in the water. But by the eve of the election, the two were neck and neck: the last public poll was released the day before the election, and it showed Kasich leading Ted by just 1 percent. It was a comeback for the ages, and it made the Ohio governor's race one of the most closely watched on election night.

Even the most skilled candidates can't overcome a true electoral wave. Ted ended up losing by just 2 percent—a devastatingly close finish in an otherwise bloodbath of a year for Democrats. He'd done everything he needed to do, and he ran significantly ahead of Democrats nationally among senior, rural, and Republican voters, but it just wasn't enough. All in all, Republicans that year picked up six US Senate seats, sixty-three(!) House seats, and six governor's seats. Even in defeat, Ted's campaign was named one of the best of the cycle by publications like *Politico* and the *Washington*

Post—in large part due to Ted's political skills and our full-frontal communications strategy against Kasich.

I had started seeing a coworker from the Corzine campaign during the 2009 race, but as is the case with most campaign relationships, it had a limited shelf life. Much like summer love, campaign love exists largely in the heat of the moment. When the adrenaline of the campaign is gone and the dust settles, it's hard to keep the chemistry going. I had some additional complications—my desire to move wherever I could find the best campaign jobs, for one. I'm also pretty sure that my close relationship with my felonious ex-boyfriend didn't help things. Jeff started serving his sentence in January 2010, soon after I began working for Strickland, and we communicated by phone and email most days. I could tell when a call was coming in from him, a blocked number with the pre-recorded message: "You are receiving a call from a federal inmate at the Manchester Correctional Facility . . ." The prison where he was serving his one-year term was a five-hour drive from Columbus. The first time I visited, I brought a Ziploc bag filled with $20 in quarters, which Jeff told me was absolutely essential.

If you've never been to a medium-security federal correctional facility, let me set the scene for you. You pull up to a gate—usually miles off the closest highway—where you're asked for your ID, and told the rules you need to follow upon entering the facility. Once inside, the guards repeat the rules to you, force you to hand over your phone, and carefully inspect your outfit (visitors aren't permitted to wear anything too tight or revealing—I was chastised for not having enough buttons fastened on my silk shirt); then you're walked through a metal detector and wanded up and down to make sure the machine hasn't missed anything. The visiting room looks like a high school cafeteria—except half the people are wearing civilian clothes and the other half are wearing prison-issued jumpsuits. You're allowed only the briefest of physical contact upon hellos and goodbyes. Once, when Jeff's hand brushed my hair instinctively, a guard growled at him menacingly: "Inmate! Don't let me catch you doing that again."

Prison food, obviously, is pretty grim, so the highlight of an inmate's week is when a visitor gets to wine and dine them with the finest fare from the prison vending machines. I put all those quarters to use and spoiled Jeff with cheeseburgers and Kraft Macaroni & Cheese that I heated up in the one microwave in the room—only visitors were allowed to operate them.

At the end of every visit, there was the option to take a photo in front of one of two crudely painted backdrops that the correctional officers would switch out depending on your preference. There was the forest scene that reminded me of my high school's production of *Into the Woods*. The other was a re-creation of Mark Twain's childhood home of Hannibal, Missouri—a picturesque river town with a steamboat in the background. It reminded me of the camera stores in Times Square when I was a kid where the owners would give tourists the option of posing in a photo with their favorite celebrity or in front of the Eiffel Tower.

I visited Jeff twice, and two visits were enough for me. Both times, I spent the entire drive back to Columbus in tears. For days afterward, I'd feel *off* and on edge. My visiting him wasn't an issue at work—I'd told the Strickland campaign that I was going to see an ex-boyfriend behind bars and that I wouldn't have access to my BlackBerry for several hours. Campaigns are generally formed in the image of their candidates, and given Ted's background of working in prisons, the visits didn't cause the slightest of concern. No one was going to judge me for spending a Sunday in a federal prison. The only time Ted ever broached the issue with me, he asked how I was handling it emotionally and if I was okay. I wasn't—it was too much for me, and I told Jeff as much. We limited our contact to emails, phone calls, and letters going forward.

A month or so after my last visit, I received a bulky letter from the Manchester Federal Correctional Institution. It was a photo of the two of us—Jeff in his green prison jumpsuit, me in a silk blouse and blazer. We looked like a couple having the time of our lives with the painted blue sky and gray steamboat smoke billowing behind us. They say a picture is worth a thousand words; in our case it was worth a thousand lies.

The Ohio race had been close, but it had still been a loss, and it stung. Looking at the margins in governor's races in the more Democratic neighboring states didn't make me feel better—in fact, they made me feel worse. In Wisconsin, the GOP candidate won by 6 percent; in Pennsylvania, the GOP candidate won by 9 percent; in Michigan, the GOP candidate won by 19 percent. Ted had outperformed the trends of the election. He should have won.

It came time to look for a new job, and it seemed only natural to go to the Democratic Governors Association. I'd worked on three of the most high stakes governor's races in the last two years. None of them were successful—something I was reminded of frequently—but I still had a feel for how they worked. There was also the allure of working for the new Chair—Maryland governor Martin O'Malley, who was being talked about as the next hot thing in Democratic politics.

I had a little more financial security by this time, but there was no question where I was going to live. When I got the DGA job offer and told my dad, he asked me: "So does this mean I get my roommate back again?" It sure did.

Sadly, it was short-lived. His second term at the SEC ended the summer after I moved back to DC. It was bittersweet for me, but it meant that he got to spend more time with my mom in New York. I still missed him every day. I missed dissecting the news with him over dinner and sharing the latest DC gossip with him—he never could get enough of the dish I had on DC politico's sex lives. I even missed his morning routine—where I'd wake at seven A.M. to the sound of his slamming the door of our apartment, only to hear him walk back in a minute or two later, exclaiming to himself, "Goddammit! Where did ah leave those GODDAMN keys?" Organizational skills were never his strength. Like father, like daughter, apparently.

I started to feel that predictable itch soon after moving back. Washington was never a place that felt like home to me. I could deal with it in small doses, months here or there, but there was always something about the town that was off-putting to me and—the kiss of death—boring. It's hard to get outside the bubble of politics there; it truly feels like a one-industry town. After you go to five parties, you realize that you've seen the same faces at every one—the same Democrats working in the Obama White House; the same Republicans working in the House and Senate; the same, young party-animal members of Congress.

I found myself getting sucked into this world where what mattered was whether your birthday was mentioned in *Politico*'s "must-read" Playbook morning newsletter—not only that, but also how prominently it was featured and who got the public credit for tipping off Playbook to it. Meanwhile, no one I knew outside DC had any idea what this "Playbook" was.

I started at the DGA in January, but by June, I started to talk with people in Obamaworld about moving elsewhere. The 2012 Obama reelection was approaching, and we all knew that it would potentially be the Super Bowl to beat all Super Bowls of presidential campaigns.

And then in September, I got an email out of the blue from one of Michelle Obama's top staffers: "Hi Liz—Could you give me a call when you have a chance?"

They were searching around for Michelle's campaign communications director. I felt slightly deflated, it wasn't exactly what I'd had in mind. I did a gut check with a couple people in Obamaworld, and one of them told me: "Not to be a downer, but I don't think you want the job. You are a hand-to-hand combat girl. This is not a hand-to-hand combat job. This is a cultivate an image by doing strategic interviews in *Glamour* and *Parade* magazine and beat back any negative stories job. I just think you would be bored. I think you want to push for a campaign war room job."

Good advice. And also—no shit! But what was I supposed to do? Turn down an interview with the First Lady's office and tank myself with Obamaworld? A few weeks later, I met with Michelle's chief of staff in the East Wing—that was cool, not gonna lie—and I came prepared after researching Michelle's initiatives and other successful First Lady initiatives on the national and state levels. I found myself simultaneously talking up how much I loved kale and her Let's Move! physical fitness cause while also praying: *Please, please, please. No. Don't pick me.*

Don't get me wrong: Michelle Obama is one of the most unique and impressive political figures in America. There is—probably with only one notable exception—no First Lady who has inhabited the White House and emerged with such a pristine brand. She is the only person who is in as equal demand as her husband on the trail. But this was admittedly a job where my strengths were weaknesses. I love to fight for my candidates. I love to roll around in the mud. I love some good trench warfare.

A couple months passed with no word from her team, and finally my electoral luck started to turn around. I'd been closely involved with the two big elections that cycle—governor's races in Kentucky and West Virginia, and miraculously, we won them both. I sent out a late-night DGA election memo crowing about the victories to the press, and it caught the Obama campaign's eye. Within a week, they reached out to see if I'd consider working on his press team. Within a month, Ben LaBolt, Obama's campaign press secretary, had offered me a gig: director of rapid response.

At its essence, campaign rapid response is the political equivalent of hand-to-hand combat. If you recognize the term, you've most likely seen the 1993 documentary I mentioned earlier, *The War Room*. Whereas prior campaign documentaries had focused almost exclusively on candidates, *The War Room* zeroed in on Clinton's top strategists during the 1992 campaign, James Carville and George Stephanopoulos. It revealed how they navigated everything from responding to attacks on Clinton's personal conduct—"Bimbo eruptions"—to exploiting President George H. W. Bush's vulnerabilities, pouncing on his fateful, broken promise: "Read my lips: no new taxes." It showed them wrestling and negotiating with the press on issues both large and small.

What made the movie so radical at the time was that it pulled back the curtain on how the sausage was made during campaigns. It showed how staff placed narratives about their opponents, and how they shaped stories long before they hit the papers or the airwaves. In some cases, they prevented stories from coming out altogether. They were critical to Clinton's success, given the sheer amount of personal baggage he had.

The War Room became an instant classic, mythologizing the role that effective rapid response played in campaigns. It made political celebrities out of Carville and Stephanopoulos.

I'm pretty sure every political operative of my generation grew up wanting to be in the war room. And here, suddenly, I wasn't just going to be in it, I would be *running* it.

The mission of our rapid response team would be the same as that of the Clinton war room—on paper, at least. But the media and technology landscape of 2012 was worlds away from 1992.

In 1992, the Clinton and Bush campaigns dealt with just one twenty-four-hour cable news channel, CNN. But by 2012, the cottage industry of twenty-four-hour news channels led by CNN, MSNBC, and Fox News played an outsized role in dictating the national political discourse. In addition to the big three, campaigns had to monitor networks like Fox Business, CNBC, and Al Gore's short-lived network, Current TV.

And then, of course, there was the internet, which was basically a nonfactor in the 1992 campaign, and no wonder—less than 6 percent of Americans had access to it at home at the time. Flash forward to 2012, and the overwhelming majority of Americans had the World Wide Web in their pocket on their smartphone. More people were getting their news from online than ever before, and social media was completely changing the rhythm of communications. The news cycle no longer came down to just what was in the morning papers and on the evening news, as it did in 1992; the onslaught of news was never ending. We didn't have the luxury of traditional deadlines—major stories were posting and running at every hour of the day. Sometimes we had just minutes to react.

My team would have to completely revamp the model of rapid response to keep up, assigning young staffers to monitor the news and social media at every hour of the day. (The most brutal shift was, of course, the overnight one; there was only so long that even an ambitious twenty-two-year-old could work those hours before burning out. So we would constantly cycle the staffers assigned to the night shift.) To give you a sense of their schedule, I'd wake up multiple times every night to check the news. And whether it was at one A.M., three A.M., or five A.M., there were always more TV and print news clips in my in-box than I could ever possibly consume.

On a visceral and primitive level, it was a job I was born for. I'm a twin in a competitive family, after all. Growing up, my twin brother messed with me daily, and in turn, I learned to mess with him back. I'd had twenty-nine years of experience being a thorn in his side, honing the art of quick comebacks, and engaging in psychological warfare. How different could it be with Mitt Romney, the likely Republican nominee?

A part of me was terrified, naturally. Outside of my time as a low-level worker on the Edwards campaign in 2004, I had no real presidential campaign experience, and I knew that I was wading into uncharted territory.

SIX

The Machine

I rolled up for the first time to the massive Obama headquarters in Chicago on a snowy Saturday evening in January 2012. It was the night of the South Carolina Republican primary.

I was bleary eyed from the twelve-hour drive, and unsure what to expect. The scene that greeted me looked like something out of a frat house, not a super-serious presidential reelection campaign. There were boxes of cold, mostly eaten deep-dish pizza (gross) and empty beer and wine bottles as far as the eye could see. Staffers were crowded around the wall of TVs in the press department, laughing and cracking raunchy jokes.

There was cause for celebration that night. Former House Speaker Newt Gingrich had come back from the political dead and shocked the establishment with a decisive twelve-point win over Romney. He delivered a defiant victory speech, letting the powers that be know he wasn't going anywhere.

The conventional wisdom heading into South Carolina was

that Romney had the nomination all but locked up. Gingrich's win threw a wrench in that plan.

It was the best possible development for the Obama campaign. The longer that Romney had to duke it out with Gingrich and former Pennsylvania senator Rick Santorum—his other big competitor—the less time he would have to focus on Obama and start the general election campaign in earnest. It felt like divine intervention.

There was one major catch, though: both Gingrich and Santorum were completely politically inept. They were political gadflies who had long been out of office. Their campaigns made the Keystone Kops look like special ops. Ronald Reagan could've dictated a primary-winning speech from the grave, and they'd have found a way to mangle it. A chimpanzee flinging a handful of feces at a map of the United States would probably have had a better voter-targeting operation than those guys did. Seriously, they were that pathetic.

It said a lot about Romney's weakness as a candidate that Santorum and Gingrich were still contenders. We were acutely aware of the fact that they had little to no ability to deal with the national press and place negative story lines about Romney, so the work of tormenting Romney fell to our campaign.

After I completed the obligatory onboarding paperwork and got my ID, I received my marching orders from LaBolt: get as many negative stories about Romney written as possible.

Sitting in the open-aired bullpen of the Obama HQ, I could hear my colleagues cold-calling reporters and pitching them with all the confidence in the world. Meanwhile, I stared at my call list, paralyzed. The pressure of working for a sitting president was intimidating. It didn't help that my makeshift desk was situated outside Jim Messina's and David Axelrod's offices. Messina—Obama's campaign manager—was responsible for a $1 billion operating budget. Axelrod, Obama's guru, was already on the Mount Rushmore of all-time greatest political strategists with people like Karl Rove, James Baker, and James Carville. What if I screwed it up? What if I'd been given this job in error, was exposed as an imposter, and was handed my walking papers?

As I started to make my way through the list of Romney embeds—the young reporters whom national news organizations devote to covering individual presidential candidates—I found the conversations awkward and brief. Essentially, "Why are you calling me?" In between each call, I'd pace around my work area, get a Diet Dr Pepper from the office vending machine, and chew down what was left of my nail beds.

Obama's reelection prospects were in serious jeopardy—no incumbent president had ever been reelected when the unemployment rate was over 7.5 percent, and the current rate was 8.3 percent. The campaign was terrified of running against an ascendant and unscathed Romney in the general election. We had to knock him down several pegs before we got to that point.

One thing was working in our favor: the Republican primary was a race to the bottom in terms of who could be most anti-immigrant or antichoice. It threatened one of Romney's strengths—that he was perceived as a social moderate.

We decided to zero in on the issue of contraception coverage under the Affordable Care Act. In January 2012, the Obama administration issued a rule that required employers to cover the costs of contraceptives, like the Pill, for female employees. The right wing was up in arms about it. Republican senators Marco Rubio and Roy Blunt introduced a bill that would allow employers to opt out of providing contraception care under the guise of "religious freedom."

It put Romney between a rock and a hard place. As governor of Massachusetts, he had ushered in a "universal" health-care law that would become the model for the Obama administration's Affordable Care Act. The law mandated that every employer cover contraception services for employees. This wasn't a position he could embrace on the national level during a Republican primary—it would torpedo him with the voters he needed to win over to secure the nomination. On the other hand, if he caved to the right wing, it would be electoral poison in a general election and reinforce his reputation as a flip-flopper. He and his campaign wisely kept mum on the topic.

I made little to no headway with the reporters on the Romney beat, even as I lobbed calls, texts, and emails into them. Any reporter worth their salt should understand the importance of birth control and the Pill—both medically and politically. So why wouldn't they just ask Romney about it?

That gets to a serious problem with modern-day political reporting. The embed system pits twentysomething reporters against hardened campaign veterans and presidential candidates, setting the stage for a Stockholm Syndrome dynamic. The young reporters, rarely acknowledged by the candidate they cover and often verbally abused by the staffers for the candidate they cover, start to feel bonded to and dependent upon them. Any scrap thrown their way, any smile, any personal acknowledgment is a gift. And so, the press corps that's devoted to covering presidential candidates can become completely unobjective. Asking a tough question might mean they end up in the back seat of the campaign plane or even kicked off. It might mean they're frozen out by the campaign staff. It's a broken system that reduces the people with the most proximity to presidential candidates to stenographers.

For four weeks, I was asked every morning and every evening by LaBolt, and Stephanie Cutter, Obama's deputy campaign manager: "Where are we on this, Lis?" The answer was always the same: nowhere. I *knew* this was a big issue, and I was prioritizing it as highly as the strategists on the Obama campaign, but I just couldn't get *any* reporters to care about it.

Then, out of the blue, I got a big break.

On the last day of February, Rush Limbaugh attacked a Georgetown Law student, Sandra Fluke, who had testified before Congress about the importance of mandatory, free contraceptive coverage. In her remarks, she invoked classmates who had faced financial hardship to afford birth control out of pocket.

"It makes her a slut, right?" Limbaugh said of Fluke. "It makes her a prostitute. She wants to be paid to have sex. She's having so much sex she can't afford contraception. She wants you and me and the taxpayers to pay her to have sex."

His comments were disgusting and depraved. He was the host of the biggest talk radio show in the country, and he was using his platform to go after a thirty-year-old, unknown health-care advocate? On the issue of the Pill, no less?

I, of all people, should not have been shocked by Limbaugh's comments. This was the guy who'd mimicked and mocked Michael J. Fox's Parkinson's symptoms with glee. But he was truly an innovator—he could always find a new low. He thought himself a clever everyman, but—once again—he was the black sheep of the family whom the Republicans would have to apologize for.

That afternoon, I was doing my usual Twitter monitoring when I saw a tweet that almost made me spit out my Diet Dr Pepper. It was from a small Columbus, Ohio, TV station—the Ohio News Network—saying that one of their reporters was preparing to sit down momentarily with Romney. Apparently, they'd secured a last-minute interview.

Romney's campaign no doubt thought that this interview was safe territory. No way would a local reporter delve into hot-button national issues like birth control coverage. What they likely didn't know, however, is that I'd spent 2010 as Ted Strickland's communications director.

During my time there, I'd struck up an occasionally contentious, but largely friendly relationship with one of the top TV reporters covering the gubernatorial race. In the waning months of the campaign, we'd meet every Monday at the Gordon Biersch next to the apartment building where we both lived. We'd watch *Monday Night Football*, eat individual kid-sized pizzas (a trick he'd taught me; they were crispier than the traditional size), drink beers, and shoot the shit.

That reporter was the one who would be sitting down shortly with Mitt Romney.

I called him. I texted him. I emailed him all the information he'd need: Romney's past positioning on the issue and details on the Senate legislation. He confirmed receipt but then went silent.

A few hours later, he tweeted: "Mitt Romney tells ONN he wld not vote 4 senate bill which wld allow employers 2 deny coverage 4 birth control." And he followed it up with a quote from Romney saying that politicians shouldn't get involved in the business of contraception.

Finally, the question we had been trying for weeks to get asked, was asked. I felt good about it, but I'll be honest: Romney's response was a letdown for the campaign. We'd been hoping for him to pander to the right. Instead, he'd said the "right" thing. "Ugh, so the general election begins" was the general sentiment in Obama HQ.

The response hit a *bit* differently on the right—it was seen as an epic betrayal of conservative values and a total "fuck you" to the Republican base. It took less than an hour for the Romney campaign to reverse course; his spokeswoman put out a statement saying that the reporter's question was "confusing" and that "Governor Romney supports the Blunt Bill because he believes in a conscience exemption in health care for religious institutions and people of faith." When the statement popped up in my email, I just about threw my phone across the room. I ran into LaBolt's office, hyperventilating: "Romney's against the mandate! He came out against it! He's against it!"

"Holy shit!" he exclaimed. "Where's Stephanie?" I asked him. "I need to tell her." As one of the few women in campaign leadership, Stephanie had been deeply invested in the idea of getting Romney on the record on this issue.

I ran across the length of the HQ to the back conference room, where she was in the middle of a budget meeting with senior Obama officials. I didn't even knock on the door before I busted in.

"Stephanie—he's against it! He's against it!"

Like everyone else in the room, Stephanie was a little taken aback at my dramatic entrance. "Who's against what?" she asked, not bothering to hide her annoyance.

"Romney! He's against the mandate. He just flip-flopped!"

"Really?" she asked, trying to process the news. After all, just an hour earlier, Romney had come out in favor of the mandate.

She quickly switched into cool professional mode. "So what are you doing about it? Have you started drafting our statement in response?"

Admittedly, I had no chill in that moment. I was like a yappy, overexcited puppy who had finally caught its tail. I hadn't thought through the next steps.

As I headed back to my desk, the HQ was buzzing audibly with excitement. I got a few high fives before I put my earphones in and banged out a statement that we would release under Cutter's name. This wasn't just my first break, it was the first time we'd successfully pitted Romney against his more conservative opponents and locked him into a far-right position.

How sweet it was. It was a dream of a twofer, playing into every possible negative perception of Romney. In the course of an afternoon, he'd managed to expose himself as both an unprincipled flip-flopper *and* a right-wing ideologue. The interview animated the national political conversation for days. It was covered on the national evening newscasts, Sunday, and late-night shows. It was written up in the biggest national publications and key swing-state newspapers. It was everywhere we needed it to be.

As soon as the ONN story became national news, the reporters who had blown off my calls and rolled their eyes at my pitches suddenly discovered my phone number and email address. They found themselves reporting off a local interview, rather than breaking the news themselves. They'd let a local TV reporter in Columbus, Ohio, drink their milkshakes.

This story speaks volumes about the media and its issues, but I'll get to that later. To me, the biggest takeaway from this episode was how much relationships matter in public and media relations. A functional relationship between a campaign and the media has to be based on trust and good faith—it *has* to be a two-way street. Cynics refer to this dynamic as "access journalism." I understand it for what it is: human nature.

Did I ever think that beers over *Monday Night Football* would help me get this big break over a year later? Absolutely not. But it built up a basic layer of trust and mutual respect. If I called the reporter about a story, it was for a good reason—it was something worth covering. I wasn't going to torch our relationship by peddling him bullshit, or by pissing on his leg and telling him it was raining.

On the flipside, I never deluded myself into thinking that a friendly, trusting relationship with the reporter would preclude him from writing a negative story about a candidate I worked for. Like me, he had a job to do. At the very least, though, it meant he would call me before he went on air and give me the opportunity to make my case. That's where the trust element is important. I can't always give on-the-record comments on stories. But I can certainly walk a reporter through sensitive and important background details that will help shape a story. The more I trust a reporter, the more I can give him.

A good flack/reporter relationship has other benefits. Reporters can't write about every story or piece of intel that comes across their desks, but they can share them with me and let me know what other people are talking about: "So and so on this campaign is peddling this research on your candidate . . . so and so at this paper is digging into this story on your opponent." Similarly, in my role, I hear *a lot* of gossip—oftentimes completely unrelated to the campaigns I'm working on. I'm never stingy about sharing the love and passing along tips.

The Romney campaign's approach to media relations was the opposite. They stonewalled reporters, refusing to cooperate on stories or provide even the most basic information. By and large, the only time the reporters would hear from the Romney team was to get yelled at. On what planet could this create a good outcome for the Romney campaign?

That's not to say that Obamaworld's approach to the media was perfect. The president would go weeks and months at a time without taking questions from the press. When he'd sit down for one-on-one interviews, he'd show as much enthusiasm for them as any normal person would for dental cleanings. His press staff was openly disdainful of the horse race and gaffe-obsessed national media. *These idiots. Don't they get what we're trying to do?*

I had a front-row seat to the sometimes troublesome nature of dealing with the news media, especially in an era when newsroom budgets were being slashed and publications were putting even more of a premium on cheap, quick hits.

At the beginning of the year, the Obama campaign handed off a massive opposition research file on Romney to a reporter at the *New York Times*. The reporter was in possession of the full kit and caboodle of the greatest hits against the presumptive Republican front-runner, including Romney's private sector record of out-sourcing jobs, laying off workers, and accruing obscene amounts of debt; and his mixed, little-talked-about four-year term as governor of Massachusetts, where he'd overseen some of the slowest job growth in the country—a record that didn't exactly support his self-styled image as an economic wizard.

The reporter never showed a lick of interest in the file, despite its obvious news value. It's unclear if he even opened it!

We didn't hear from him for weeks as he filed forgettable puff piece after forgettable puff piece about Romney. That was until he read a story in *Politico*: "Romney's 4-Car Fantasy Home."

> *At Mitt Romney's proposed California beach house, the cars will have their own separate elevator.*
>
> *There's also a planned outdoor shower and a 3,600-square-foot basement—a room with more floor space than the existing home's entire living quarters. . . .*
>
> *A project this ambitious comes with another feature you don't always find with the typical fixer-upper: its own lobbyist, hired by Romney to push the plan through the approval process.*

The *Times* reporter obviously put two and two together that *Politico* hadn't come by this information by posting a reporter at the San Diego courthouse. He was apoplectic when he called up LaBolt: "*How* could you give that story to *Politico—that rag!*"

"What are you talking about," LaBolt shot back. "We gave you *everything*! *Every* serious critique of the Romney record—you used *nothing*! Now you want some clickbaity *trash*?" LaBolt never held back on the phone with reporters or in person with staff. He led every morning meeting with an Al Pacino–esque monologue. Sometimes it was the inspiring Pacino from *Any Given Sunday*, other times the self-righteous Pacino from *Scent of a Woman*, and during the dark days the murder everyone and their children Pacino from *The Godfather*. He'd worked in New Hampshire for Dean, in Ohio for Sherrod Brown, in the White House for Obama, in Chicago for Rahm Emanuel, and then come back to be Obama's top spokesperson. He knew politics at every level. He didn't have time for reporters' gripes or staffers' excuses. He went through assistants faster than Naomi Campbell. He called out bullshit when he saw it.

LaBolt was right—the story was clickbaity. It provided a compelling visual for Romney's exorbitant wealth. It also reinforced his complete tone deafness as a politician. Here he was plotting a run for president and he didn't even think twice about hiring a lobbyist to file public documents requesting permission to build a car elevator?

Our video team went to town producing online videos modeled off the old TV show *Lifestyles of the Rich and Famous*, where we walked voters through the perks of being Mitt Romney. It got under the Romney campaign's skin, but still, *come on*. Watergate, this was not. No one was winning a Pulitzer for this story.

There was much more important information out there about Romney—information that was far more revealing about the type of president he would be. It would require more work to report on, sure, but in terms of the national public interest, what was more relevant: that Romney had gotten rich by destroying companies, laying off thousands of jobs, and avoiding his fair share of taxes? Or that he was building a luxe vacation home that was more relatable in Dubai than in Dublin, Ohio?

It felt like campaign reporters wanted the prestige of working at marquee publications without doing any of the hard, gumshoe reporting of their predecessors. In earlier presidential cycles, the reporters that covered campaigns were hardened veterans or, at the very least, people who had spent several years cutting their teeth in less-glamorous gigs in state capitols and city halls across the country. There's not much glory in filing dozens of public information requests to get to the bottom of sketchy construction contracts in Youngstown, Ohio, or taking two years to report on the self-dealing practices of a small-town sheriff in Nevada. But that experience teaches a reporter a lot about politics—how to find the "news" and how to smell through the bullshit and spin fed to them by politicians and their flacks.

The burden here can't be placed solely on the reporters. National news organizations made a conscious decision to change their own paradigm of campaign coverage. They no longer wanted older, more experienced journalists who would doggedly and thoughtfully report on a couple stories a week, tops. They wanted twentysomethings who could work 24/7 on the cheap, transcribe speeches and press conferences quickly, and churn out multiple "stories" a day. News organizations got what they paid for, ending up with greener, more impressionable reporters who were unlikely and unable to speak truth to power.

In part, it was a response to the changing nature of information and media consumption underfoot. Obama's 2012 campaign was at the forefront of this brave new world.

After the 2008 race, the Obama campaign was held up as the gold standard of how to master the use of digital and new technologies in campaigns. Four years later, it was a completely different beast.

In 2008, the campaign's digital department was walled off

from the communications department, focused entirely on online organizing and fundraising. By 2012, it was fully integrated into every aspect of the campaign—including two designated staffers on my rapid response team—and we completely dominated Romney across every platform in terms of followers and engagement. Obama had 20M followers on Twitter to Romney's 1.5M; 29M followers on Facebook to Romney's 8M; 233K subscribers on YouTube to Romney's 21K; and 1.4M followers on Instagram to Romney's 38K.

The digital gap was stunning, and the campaign had built the tools to exploit the shit out of it. Obama's reelection campaign was the first to have its own in-house production team. We could turn around slick videos, graphics, and memes in record time and distribute them to our email list, social media platforms, and the news media.

On a daily basis, you'd see campaign officials parked in front of the HQ camera that provided a live feed to news stations. Except instead of doing just interviews, they were spending half their time recording their own videos—videos that mimicked the look and feel of news reports, but that typically had much higher production values.

It was a game changer. If we thought that the media was ignoring an important story, we didn't just sit on our hands and whine about it, we'd create our own news and bypass them. We produced short, snappy explainers of complex policy issues. We translated long-form, investigative print stories into digestible, sharable video content. We offered real-time responses to Republican attacks, rather than waiting for the chance on a Sunday show days later. A May video starring Stephanie Cutter, Obama's deputy campaign manager, titled, "Get the Facts on Mitt Romney, Big Oil, and the Koch Brothers," generated more than eight hundred thousand views. The viral nature of these videos forced the media to cover stories that they'd normally ignore. We created our own echo chamber.

Cutter's direct-to-camera video about the Koch brothers generated infinitely more views than any direct-to-camera video we put out from the First Lady. It more than doubled the views of Jennifer Lopez's direct-to-camera endorsement of Obama that year, her first presidential endorsement ever.

Why did it get such a big grassroots response? We were the first campaign to really understand the power of parasocial relationships between supporters and the candidate and his staff. À la *The War Room*, we pulled back the curtain on one of the most opaque professions out there, demystifying and personalizing it for the general public. The more intimately people got to know the campaign, the more invested they were in the fight. On debate nights, they felt like they were *in* the war room with us because of the access that we gave them. It made them more likely to chime in online and donate money.

That wasn't the only medium that was changing that cycle. The rise of Twitter was completely reshaping how both campaigns and the media operated. It was another total game changer, for both good and bad.

In 2008, Twitter existed, but it was largely a novelty. I set up a Twitter account that spring. During the McAuliffe and Corzine campaigns in 2009, I was the only press staffer regularly using it as a communications tool—I saw it as a fast-and-easy way to pump out news clips into the ether. It couldn't hurt, right? I had just over a thousand Twitter followers when I started on the Obama campaign. The fact that that seemed like a relatively large number speaks to what a newish platform it was.

In 2012, Twitter became a force of its own. It changed the way we operated on the campaign. Every staffer was told to open a

Twitter account when they were hired. It became a regular tool for the communications staff, one that allowed us to shape the news and get out spin in real time. Up to that point it seemed unfathomable that a 140-character post could change a news cycle, but it did.

Just as an example: one afternoon while trawling Romney's event coverage, I noticed a quote in a *Milwaukee Journal Sentinel* story about how Romney told voters a "humorous" anecdote about his dad closing a factory in Michigan. It was vintage Romney, awkward and out of touch. I hadn't seen the quote in any other coverage or on Twitter, so I tweeted it out to my now-few thousand followers. From there it took on a life of its own—the *New York Times* and viral online outlets like BuzzFeed posted stories about it. It was all over cable news that night, including Rachel Maddow's show on MSNBC. A single tweet disrupted the presidential news cycle, throwing Romney's campaign off message for days. Without Twitter, Romney's comments would probably have faded away.

Throughout 2012, I listened to every single Romney event as they took place. I was terrified of missing a single utterance. To block out the noise and distractions of our bullpen of an office, I had earphones in at all times. I wore sunglasses at my desk, because they helped cut out the glare and commotion. They also sent an indisputable message to my coworkers: "Do not *even try* to strike up a conversation with me."

My singular focus and intensity became a source of amusement to my coworkers—they would take turns coming up behind me and tapping on my shoulder, a practical joke that would usually result in my jumping out of my skin, knocking over Diet Dr Peppers, and cussing out the perpetrator loudly. By the end of the campaign, my deputy director of rapid response, Danny Kanner, had adopted the same practice, wearing his earphones and Ray-Ban aviators at his desk. We must have looked ridiculous, completely silent as we conducted our conversations over GChat, even though we were literally sitting across a desk from each other. We were like the campaign equivalent of the Blues Brothers—but it worked for us.

Twitter brought more transparency to the behind-the-scenes machinations of campaigns. In the past, the distasteful art of spinning was usually reserved for one-on-one phone conversations or email exchanges. Now, our press shop was doing it in full public view, posting Romney's policy inconsistencies, mocking his awkward public statements—"This feels good, being back in Michigan. The trees are the right height"—and highlighting gaffes he made in real time, forcing his team to respond as soon as his event concluded.

Twitter also changed the way the media worked. It reduced the stranglehold that the big, legacy media organizations had over the news cycle. Peter Hamby, a CNN political reporter, captured the dynamic in his post-2012 paper for the Shorenstein Center at Harvard: "Did Twitter Kill the Boys on the Bus?"

> *Placing a story on the front page of* The New York Times *print edition didn't really matter so much in 2012 outside the confines of the "Morning Joe" green room. . . .*
>
> *"A link is a link," said Matt Rhoades, Romney's campaign manager. "I've said this a million times. I used to say it going back to, like, 2004. A link is a link, dude."*
>
> *It might be a link to* The New York Times *or* Politico *or* RedState *or* ThinkProgress—*an online story, no matter how biased or thinly reported, had a URL that could be peddled to other news outlets, linked to the Drudge Report, and, most importantly, pumped directly into the Twitter feeding frenzy where influencers lived.*

Just think about the example of Romney's contraception comments: a question and tweet from a relatively obscure regional TV network was enough to drive local and national media coverage. The news trickled up rather than down.

It had a democratizing effect, for the good, but it also created a lowest-common-denominator phenomenon among the media. No reporter wanted to get scooped on anything, and so they'd tweet out every detail of what they saw, from what a candidate said to what they wore and ate. There was no tidbit too insignificant to share. It led to inevitable mistakes: reporters would misquote and misrepresent the candidates and their supporters, and their tweet would travel halfway around the internet before it was deleted or corrected.

If cable news had eroded the thoughtfulness of news coverage, Twitter demolished it. In a twenty-four-hour, cable-news-fueled news cycle, where every hour every show needs a new angle, it's "easier" to focus on things like polls and minor gaffes. You can turn around a story or a TV package on how Mitt Romney responded to a poll faster than you can develop a piece on his complex business dealings. Some of the rationale for round-the-clock political coverage is that it will somehow hold politicians more accountable. Perversely, it's had the opposite effect. It's reduced politics to a spectator sport, where meaningless minute-by-minute developments are blown out of proportion, turning the smallest stories into the biggest ones.

Our campaign never lost perspective on Twitter, and just who we were reaching with it. We viewed it as a tool to influence insiders and news coverage, not something that reached the masses or reflected "real life" in any way. Just 15 percent of all American adults were on Twitter in 2012 and only 2 percent of American adults checked Twitter regularly. Drawing any broad conclusions from Twitter would have been futile.

Too many presidential campaigns today missed that memo. In 2019 and 2020, Twitter played an outsized role in the Democratic presidential primaries among too many campaigns—despite the fact that just 22 percent of Americans were estimated to have a Twitter account.

In October 2020, Pew Research Center dug into the numbers to show just how skewed a perspective the platform represented:

> *A small share of highly active users, most of whom are Democrats, produce the vast majority of tweets. The Center's analysis finds that just 10 percent of users produced 92 percent of all tweets from U.S. adults since last November, and that 69 percent of these highly prolific users identify as Democrats or Democratic-leaning independents.*

It found that the Democrats on Twitter were younger, more liberal, and more likely to have a college or postcollege degree than non–Twitter users. Basically, the study found that Twitter is not representative of the electorate.

Early in the 2020 cycle, I remember a reporter confiding in me how an opposing campaign had told him that their goal every day was to "win Twitter." Yikes, no wonder they never even made it to Iowa.

Events in 2012 demonstrated how even the most disciplined of campaigns—Obama's, in this instance—could lose control of its supporters and the narrative. It was a dynamic that was unquestionably amplified by Twitter.

In May 2012, our campaign launched an anti-Romney offensive with an ad featuring steelworkers who'd been laid off from their jobs after their Kansas City plant was bought out by Bain

Capital during Romney's tenure as its CEO. It wasn't your normal thirty-second ad—it ran two minutes and was intended to signal to voters and elites that we weren't going to let the election be a referendum on the president's economic record; we were going to make it one on *Romney's*. The language in the ad was blunt. One steelworker compared Bain to "a vampire. They came in and sucked the life out of us." Another said that watching Bain destroy his plant "was like watching an old friend bleed to death."

Hard-hitting? Sure. But c'mon. Politics ain't beanbag. We weren't going to give Romney a free pass on his private sector record, and these were workers describing their own feelings about Romney.

The Sunday after we put out the ad, my team was the first to alert the higher-ups at the Obama campaign that we had a problem. One of our roles in rapid response was to monitor all the Sunday shows on TV—the venerated weeklies like *Meet the Press, This Week, Face the Nation,* and *Fox News Sunday*—that, while declining in influence, still set the week-ahead political agenda. We made a point of trying to stack the shows with favorable surrogates—elected officials who would carry the party line for us. We'd speak with them in advance, give them talking points, and do everything in our power to make sure that they were loaded for bear when they hit the set.

So I was blindsided when a member of my team flagged a panel on *Meet the Press.* Cory Booker, the then-mayor of Newark and a close Obama ally, was asked about the Bain ad by the then–*Meet the Press* host David Gregory, and he didn't hold back his criticism. He said he was "very uncomfortable" with the ad and that he found it "nauseating"—even comparing it to Republican attacks against Obama for worshipping at Reverend Jeremiah Wright's church. "If you look at the totality of Bain Capital's record," Booker said, "they've done a lot to support businesses, to grow businesses."

Oh shit. This is bad. This is really, really bad. That was all I could think to myself as I waited for my team to clip the video and circulate it more broadly. There was no question in my mind that this would likely dominate the news for days, if not weeks, to come. After all, Booker was one of Obama's top supporters—and here he was saying that our ads made him want to vomit.

It was a PR disaster—"The surrogate from hell" was how MSNBC's Steve Kornacki described Booker's appearance. Bain was supposed to be our ace in the hole against Romney—it was by far the most disqualifying attack that we had polled. Suddenly, though, it had become the bane of our existence.

Everywhere we turned, Booker's comments were being played and replayed, on TV news and in Republican ads. To make matters worse, other Democrats soon jumped into the fray to attack the ad, including then–Massachusetts governor Deval Patrick: "I think Bain is a perfectly fine company. They've got a role in the private economy." And Bill Clinton: "I think [Romney] had a good business career . . . I don't think that we ought to get into the position where we say this is bad work." The comments from these business-friendly Democrats drove the narrative in the Beltway that the ad campaign had backfired disastrously. It was hard to find a pundit or columnist or TV host who didn't share this opinion.

It was a great illustration of the disconnect that exists between national political and media elites and voters outside of the Acela Corridor. A couple weeks later, the Obama campaign got back new internal polling that tested the impact of the controversial Bain ad. It showed that Obama was opening up a lead over Romney on key metrics such as whom voters trusted most to manage the economy, who would "fight for people like you," and who would build an economy that strengthens the middle class.

We won the internal party debate once the numbers moved our way. Romney's business dealings were fair game. The Democrats who thought we had crossed a line were being overly precious.

Romney's business career was fair game, for sure. Other things, like his membership in the Church of Jesus Christ of Latter-day Saints, were completely off the table. In 2008, both Obama and Romney faced ugly attacks for their religious affiliations. In Obama's case, it was his ties to the aforementioned Wright, a controversial Chicago-based pastor who'd delivered anti-American sermons. Romney, on the other hand, was targeted with anonymous "push polls" in Iowa about his Mormonism, and one of the top GOP candidates that year, Mike Huckabee, raised Romney's faith on the campaign trail, asking a *New York Times* reporter who was profiling him, "Don't Mormons believe that Jesus and the devil are brothers?" I'm *sure* he was just curious.

Attacks on Romney's faith were forbidden within the Obama campaign—no ifs, ands, or buts about it. The campaign wasn't interested in playing dirty, but we also knew that any such attacks could backfire and transform the wooden, unlikable Romney into a victim and a sympathetic figure. It was a dynamic that I had internalized after the 2009 New Jersey gubernatorial race.

Our campaign was remarkably disciplined. But the reality of a presidential campaign is that outside of the candidates and their staffs, there's an ecosystem of surrogates, supporters, and hangers-on that you can never control. I can't even count the number of times that we had to disavow controversial, off-message statements from gadflies on cable TV. But not every problematic statement came from a clout-chasing nobody. Sometimes they came from prominent members of the party.

"FML," I GChatted a coworker of mine when I saw the Daily Beast headline land in my email in-box: "Brian Schweitzer: Mitt Romney's 'Family Came from a Polygamy Commune in Mexico.'" In an interview, Schweitzer—the eminently quotable former Democratic governor of Montana—invoked "polygamy" no fewer than seven times in reference to Romney.

His comments weren't just unhelpful; they were our worst nightmare. In the spring of 2016, President Obama was still fighting off the ridiculous birther conspiracy theory, and our campaign had staked out a clear high ground against its often racist backers on the right. Comments like these would only reduce our moral standing.

As soon as the story was circulated, I was called into Cutter's office and told, as someone who'd worked with Schweitzer in the past at the DGA, to "handle" it. I called the governor's top political staffer to ask what the hell had happened. When I got him on the phone, he tried to minimize the comments and the governor's intent, explaining hurriedly that the interview was supposed to be on a different subject, but that it had gotten off topic. *Well, no shit.* I gave them the courtesy heads-up that we'd be denouncing the comments, and—if possible—that we'd like them to apologize for them as well.

The backlash was swift. Never one for rapid response, Romney was crying foul on cable TV the next day. Republicans, Democrats, and media personalities like Jon Stewart slammed the comments. It didn't help that Schweitzer, through aides, refused to walk them back: "The governor believes exactly what he said: that Romney is in a pickle. Romney will probably not choose to highlight his own family's connection to Mexico as a way of reaching out to Hispanics, because that history involves a polygamy colony, which is something that Romney doesn't like to discuss."

The story dragged out for a few days. It was on a whole different level than the Bain debate, which was legitimate. It was *juicy* and *scandalous.* It was a helluva lot more exciting for reporters than covering policy debates. Just as we thought it was dying down, I got word that Schweitzer was scheduled to be on Anderson Cooper's CNN show that evening. When I informed the campaign, they had an aneurysm. I called Schweitzer's adviser to ask what they were thinking, and he texted me back that he was on a tarmac and would call me when he landed.

I watched the clock tick as the hit approached. A couple hours later, I saw my desk line light up with a call from Schweitzer's adviser. When I picked up the phone, however, I heard Governor Schweitzer's signature twang.

"Hey cowgirl! So, I understand you guys aren't too happy with my comments about the polygamy commune, huh."

"That is an accurate assessment," I told him.

"Look, someone needed to poke the tiger on Romney. I know you had to disavow me, and I'm happy to have you do it, but someone needed to poke the tiger."

"Consider the tiger poked," I told him. "So . . . I see you're going on CNN tonight. Anything we should be aware of?"

I was trying to tread lightly, knowing that he didn't like to be "controlled." One of the first times I'd interacted with him was prior to a DGA press conference in Salt Lake City, Utah. When I'd handed over to him a set of suggested talking points for the presser, he had mocked the gesture. "So, I guess this is what you want me to say, huh?" And flipped through them dismissively before he went up to the podium and did his shtick as a western governor with a big belt buckle, and bigger personality.

"I said what I had to say on that topic," he responded. "I'm not going to take it back, but I don't need to say it twice."

The call ended, and I walked into Stephanie's office to inform her of the latest developments.

"So? Is he going to apologize or what?" she huffed.

It was a serious question. And this was serious business. Schweitzer's comments about Romney were offensive, off brand (not for Schweitzer, but for the Obama campaign), and supremely unhelpful. Still, I found myself suppressing a giggle as I recounted the conversation.

"Well, I don't think he's going to apologize. He said he had poked the tiger and that he wouldn't poke the tiger again."

Even she had a hard time keeping a straight face as I repeated the governor's comments to her: "What does that even mean?"

I told her that it would be fine and that I'd monitor it.

An hour later when Schweitzer appeared on Anderson Cooper's show, he was matched up with the religious right leader Ralph Reed. Both Reed and Cooper tried to bait Schweitzer into repeating his comments. It must have been the most frustrating interview of Cooper's life.

Schweitzer kept his word, giving one of the more evasive interviews I'd ever seen. He added no fuel to the fire and the controversy died down soon after. (It wasn't the end of our issue with Democrats or left-leaning figures attacking Romney's religion, of course. A month later, Bill Maher, who had given $1 million to a Super PAC supporting Obama's reelection, referred to Mormonism as a "cult." The same fury ensued.)

The CNN interview—and Schweitzer's dogged refusal to repeat his initial remarks—made for a humorous bookend to an especially unpleasant moment in the campaign. I think a lot of things are fair game in politics; candidates' religious beliefs shouldn't be. If we were asking people not to judge Obama based on his affiliation with Reverend Wright, it would have been the height of hypocrisy to ask voters to judge Romney for being Mormon. We had enough ammo against him; who cared how he worshipped?

The Schweitzer affair illustrated the absurd outrage cycle of politics. One former Democratic governor made some unfortunate comments. Republicans ran wild with them. And jaundiced political observers wondered aloud whether this was a "trial balloon" from our campaign—an attempt to test out whether voters would be turned off by Romney's religion.

Nothing could have been further from the truth. We'd absorbed

Schweitzer's comments with abject horror. What seemed like a moment straight out of *House of Cards* was in reality an outtake from *VEEP*.

I certainly don't blame anyone for ascribing the worst possible motives to politicians; it's well deserved. But sometimes politicians are just like the rest of us: they fuck up. They say idiotic and unhelpful things with no strategic end goal. Even the most skilled of them—if caught off guard at the wrong moment—can exhibit the same tact level as a drunk uncle at Thanksgiving. That doesn't excuse comments like Schweitzer's, but it also doesn't mean that there's always some grand plan behind them.

Onward, we marched. And with a crystal-clear directive to anyone associated with the campaign: attack Romney at will—for his corporate background, his sketchy record as governor of Massachusetts, his finger-in-the-wind political style. But under no circumstances should you attack his faith.

As the campaign was entering the home stretch—the period from Labor Day to Election Day—I was confronting a personal-life reckoning moment of my own: my twin brother, Angus, got married. We'd spent eight months in our mother's womb together. I was born thirteen minutes before him, and for the rest of our lives I'd quote the line from the Arnold Schwarzenegger–Danny DeVito movie *Twins*, that he was "the crap that was left over." I would've used a more clever line if Hollywood had ever produced a more creative movie about twins, but to this day that's all I've had to work with.

It was weird seeing Angus and my friends from high school— even the rowdy ones—engaged or pregnant. Over the last four years, I'd lived in seven different states, and I had at least as many boyfriends to show for it, including my forgettable date that night, who got mad when I told him I wanted to go to the afterparty for my *twin brother's* wedding.

I saw Angus's happiness. I saw how happy that happiness made my siblings and my parents. I saw my friends' happiness with their spouses and fiancés. And I shared that happiness, truly I did, even if I didn't experience that magical wedding moment where some song temporarily convinces you that the random guy who asks you to dance is in fact your soul mate. I'd made a point of putting my phone in my bag and not checking it obsessively, but by the time midnight arrived, what I wanted more than love was to know *what the fuck was going on with Benghazi.*

On the Obama campaign, we had dealt with points of crisis: the mass shooting in Aurora, Colorado, where a gunman killed twelve people at a midnight screening of one of the Batman movies. It was the deadliest mass shooting since Columbine, and, looking back, it's chilling to think how abnormal it seemed at the time. Both the Romney and Obama campaigns effectively shut down for a week: we pulled down ads, we didn't tweet negative political stuff, we held back on statements to the press and emails to supporters. A month later, a gunman killed six people in an attack on a Sikh temple in Wisconsin. We went quiet for a day or two and then went back to work.

And then, there was Benghazi. On September 11, 2012, the American embassy in Benghazi, Libya, was stormed and four Americans were killed, including the US ambassador to Libya, Chris Stevens. The facts weren't even out before Romney released a statement calling the Obama administration "disgraceful" and saying that they "sympathize[d] with those who waged the attacks."

Back in the days before Donald Trump won election, there was this vaunted idea of the "commander in chief" test—could presidential candidates rise to the occasion of an international crisis? In some cases, the test was administered on Sunday shows, where people like John Edwards would flail at basic questions about foreign relations. This was a real-time, very serious test for Mitt Romney, and he couldn't have failed it harder. Romney would pull off mini-comebacks after Benghazi (see: the first presidential debate), but his shaky press conference the morning after he'd put out his ill-advised statement felt like the moment when the wheels started to come off the bus. He looked angry and accused Obama of "apologizing" for terrorism. The media split screened it with Obama's lofty speech from the Rose Garden, where he honored the Americans who'd lost their lives in the attack.

It was the only weekend I took off from the campaign and the only weekend other than the aftermath of the Aurora shooting when our campaign's press operation went dark. Nevertheless, I couldn't mentally tear myself away from the news. I was glad Angus had settled down and found his one true love, but it was clear mine was still work.

Maybe it was the all-consuming nature of my job. Maybe it was the mediocrity of my date. But I can still remember then–US ambassador Susan Rice's lines on the Sunday shows (the appearances had been disastrous). I couldn't for the life of me tell you what song my twin brother and his wife chose for their first dance.

SEVEN

Blurred Lines

Obama would go on to win the presidential election fairly handily—he carried every state he'd won in 2008 except for Indiana and North Carolina. The afternoon after the election, he popped by the campaign headquarters to surprise the hundreds of sleep-deprived and hungover staffers who were cleaning out their desks. Over the years, Obama had developed a reputation for being a little too cool for school—aloof and detached at times. But just a few minutes into his speech thanking us, he choked up and paused to wipe away the tear that was running down his right cheek. After he wrapped his speech, he stayed for hours, hugging every single staffer in the office, looking each one of them in the eye to say "Thank you." He was a graceful and classy leader and—I'd be remiss to omit—a world-class hugger.

Whereas on the 2008 campaign, staffers had some assurance of getting a job in his administration, it was much more uncertain in

2012. Obviously, all the plum jobs were already taken by Obama people.

I hadn't shared my plans with any of my colleagues—I didn't want to jinx anything; I didn't want to seem cocky. But when I woke up the day after the election, I knew what I'd be doing. In August, at the Republican National Convention in Tampa, I'd talked with Martin O'Malley and the executive director of the DGA, Colm O'Comartun. O'Malley's term as DGA chair was coming to an end, and he wanted to run for president. The DGA wanted a steady communications hand on board. We did a gentleman's handshake and agreed that if I could present a plan for both, I'd consult for them.

The term "political consultant" is sort of fraught. The stereotype is a way-overpaid adviser—usually male, usually older. I wasn't that. I had just turned thirty. But I knew what animated me in life: it was politics and political campaigns. Between August and November, when—finally—I'd worked on a winning campaign again, my conviction intensified.

To outsiders, it would seem like the natural next step would be to parlay my campaign work into securing an administration job. That was never a consideration for me. Short of being White House press secretary or a spokesperson at a handful of departments like State or Justice, there were no jobs that stood out as especially enticing. The prospect of ending up in a windowless basement office at the Environmental Protection Agency, churning out press releases about CAFE (Corporate Average Fuel Economy) standards and how the Obamas' dogs were celebrating Earth Day, was as appealing to me as a career in the legal industry—just because it was something that I was expected to do didn't mean I wanted to do it.

I'd loved my time working with O'Malley. We spent a lot of time together when he was the chair of the Democratic Governors

Association; I'd prep him for interviews, travel with him, and hit the occasional bar or two. I'd gotten to know his family as well and adored them.

My official meeting with O'Malley regarding my hiring took place right after Thanksgiving. I put together a pitch document for our four P.M. meeting in Annapolis. We met at the Galway Pub, an Irish bar that was a five-minute walk down the hill from the state capitol. As we sat down for our meeting, O'Malley ordered water but changed his mind after I ordered a Guinness.

That afternoon, as I'd driven to the pub, news had started to break about a school shooting in Connecticut. I'd largely tuned it out. But as we sat in the Galway, we watched the bartenders turn up the volume on the TVs. Dozens of young children had been shot and killed at Sandy Hook Elementary School, the casualties were unlike any from a school shooting in American history. After I walked through my proposal, O'Malley folded the multipage document in half and placed it in the sleeve of his iPad. "Holy shit, this is really bad." Our catch-up quickly came to an end—he wanted to monitor the situation more closely and be with his family.

I served as an outside adviser to O'Malley and his administration—making sure that their governmental agenda aligned with his political one. He'd already decided that one of the issues he would expend political capital on during the 2013 legislative session was abolishing the death penalty. It was a politically toxic issue—the majority of Marylanders opposed getting rid of it. No matter how many times his advisers informed him of that fact, he never wavered on his commitment to abolish it: he was a social justice Jesuit to his core, and he believed that the death penalty, so long as it was legal, would be a blight on his state. And since he was going to go all-out against the death penalty, why not take on the sacred cow of gun rights?

Thirty-five days after our afternoon meeting in Annapolis, he was the first governor to announce a truly sweeping gun safety package. The announcement was made on the front page of the *Washington Post*. He would be pursuing bans on high-capacity magazines and assault weapons, along with mandatory fingerprinting and licensing for anyone purchasing a handgun. It was radical in how comprehensive it was, and the Second Amendment dead enders came out in full force against it—holding one of the largest demonstrations in modern Maryland political history outside the capitol. O'Malley even met opposition from legislators in his own party. None of it mattered to him. Even in the early 2000s, when most Democrats ran as fast and far away from gun control as they could, he had been an ardent supporter. He'd seen the human toll that gun violence took on his city of Baltimore, and he was also hardheaded in his convictions. In April, he signed the legislation that he'd proposed into law, with very few modifications. He also enacted a ban on the death penalty in Maryland. None of these actions helped make him particularly popular, but it was critical to him to do the right thing.

The Maryland legislative session is relatively short—just ninety days every year. Once the General Assembly wraps and the governor signs that year's legislation, there's a steep drop-off in media coverage and drama. I was still getting a paycheck but didn't have a ton of work on my plate. And again, I found myself stuck in DC, going to the same parties with the same people, having the same conversations over and over again.

That was the backdrop to the call I received out of the blue in July from Hari Sevugan, a friend and former colleague. Hari and I met in 2010 when I was working for Ted, and Hari was the Democratic National Committee's press secretary. Whenever I wanted to get more national attention for something that was happening in our race, Hari was the person I'd call for help. Now, the tables had turned. He was working as a senior adviser on Eliot Spitzer's splashy comeback campaign for New York City comptroller. He cut to the chase: "Would you consider working for him?" (Six years later, we'd have this conversation in reverse, when I'd call him to see if he'd join Pete Buttigieg's presidential campaign. Yes, politics *is* a very small world.)

By now, pretty much everyone is familiar with Eliot's story. Eliot had been a *star* as AG. He earned the moniker Sheriff of Wall Street by taking down people like Hank Greenberg, the CEO of AIG, who admitted to committing at least $500 million in accounting fraud. When "E" as I liked to call him ascended to the role of governor of New York, he was frequently mentioned as a presidential candidate. "First Jewish President?" was a typical headline in the papers those days.

Eliot had flown too close to the sun, and—like everyone else—had his own flaws and demons. He shocked just about everyone when, a little over one year into his term as governor, he pulled together a hastily announced press conference to confess that as attorney general and governor, he had solicited prostitutes. You can recover from a lot of sins in politics, but rank hypocrisy is not one of them, especially when your enemies are the most monied and powerful people on Wall Street. He resigned from office and went from golden boy to a punch line in under twenty-four hours.

"It's going to be a crazy campaign," Hari told me. "But we have a chance to do something big here—rehabilitate this guy's career. And you came to mind, because you're crazy enough to do it."

I mulled the offer over for a few days. I sought the counsel of former colleagues and mentors. The consensus was nearly universal: it's a big risk; don't do it. So naturally, I packed an overnight bag and bought an Amtrak ticket to New York City. What could go wrong?

I quickly learned the answer.

My first day on the job, the *New York Daily News* broke news of my hire, noting: "Smith isn't just not afraid of pushing back on behalf of her candidate; she's—yes, total reference to the 2006 Spitzer bio by Brooke Masters—known for, more often than not, spoiling for a fight." On my third day, right-wing site Breitbart picked up the *Daily News* story and speculated that I was hired "due to growing rumors that Spitzer is engaged in an affair with a married woman."

A ridiculous assertion, it seemed, until I was waved into a "sensitive" meeting with Eliot and a few others—his campaign manager, Hari, and his longtime PR representative. They'd been having a hushed conversation for the last fifteen minutes, something I was well aware of due to the intimate nature of Eliot's real estate office. The headquarters of Spitzer Enterprises was on the twenty-third floor of the Crown Building on Fifth Avenue. It was some of the most coveted real estate in the Big Apple—from his office you could look out onto the immense expanse of Central Park. Just across the street was the Tiffany's flagship store that Audrey Hepburn had turned into one of the most recognizable shopping destinations in the world. All that glamour aside, the campaign staff was given just two cramped offices to work out of. Most days, you'd see six of us piled into the small office next to Eliot's. Privacy was virtually impossible.

I was beckoned to the file room where they were having their conversation, and I took a seat on the floor. The mood was tense, and I soon learned why. The *New York Post* was threatening to run with rumors that Eliot had a new girlfriend, and their deadline was in a matter of minutes. It was an open secret that Eliot was separated from his wife, but the press hadn't yet nailed down the story.

Eliot, whom I'd first met the day prior and found more amiable and humble than his public persona, had made up his mind. The plan? Deny, deny, deny. Gone was yesterday's friendly demeanor; it was clear that this was an order, not a suggestion.

I picked up on the internal dynamics quickly that first week. His PR representative gave off strong yes-woman vibes. She was resigned to doing exactly as he said. I had no such compunction—a yes-woman I am not.

I didn't know Eliot on a personal level, and I certainly didn't have the full background on this particular story. But—like the rest of America and the two other political aides in the room—I'd read enough to have my suspicions. So I laid out the options on how we could respond and how they'd likely play out in the press.

The first option—a blanket denial—might seem like the path of least resistance, I explained, but would expose Eliot to further scrutiny. Whether or not he had a "girlfriend," as the *Post* was alleging, an unambiguous denial of the charge could open up the definition of "girlfriend" and invite the tabloid press to dig deeper into his personal life.

Think of Gary Hart during his 1988 campaign, I told them. Faced with questions about his personal life, Hart had offered fulsome denials of any extramarital relationships and encouraged reporters to "put a tail on" him. His campaign ended, effectively, when a young woman—not his wife—was caught leaving his Capitol Hill abode. The scandal burst wide open when photos were published of said young woman sitting on Hart's lap on a yacht aptly named *Monkey Business*.

I offered up a second, preferable option to Eliot's: stop commenting on allegations about his personal life. Starve the narrative. If the tabloids had a solid story, they'd run it.

The room went silent. Even with his political power diminished, Eliot cut a daunting figure. People didn't usually challenge him or tell him things he didn't want to hear. But when I did, he was unfazed. He paused for a second to think it over. He asked if a "no comment" would be viewed as a confirmation of what the *New York Post* thought they had. Then he had me run through the different scenarios for him once again.

Satisfied with my responses, he took a moment. "Lis is right. Let's go with no comment." And so we did and braced ourselves for an ugly story in the *New York Post*.

When I woke up early the next morning, there was nothing on the *Post's* website. Nothing in the paper. There was no story—all because Eliot refused to comment. Later, he told me it was the best decision he'd made during the campaign.

There's this stereotype of politicians—of most people in powerful positions, really—that they listen only to the things they want to hear. That anything other than a "yes, sir/ma'am" will put you in bad favor, reduce your influence, maybe even get you fired. My experience has been the exact opposite: the more powerful people are, the more they actually want to hear the word "no" and be challenged, as long it's with a sound argument. Strong people recognize strength and respect it even in people with whom they vehemently disagree. More concisely, game recognizes game.

For a couple days afterward, the *Post* and *Daily News* continued to hound him. They wanted to be the first to break the news that Eliot's marriage was over. We didn't give them any quarter, and they moved on.

For a while, the strategy worked. In the history of New York City, I doubt you could find someone more overqualified for the position of comptroller than Eliot. In interviews, he would run circles around reporters who asked him what he wanted to do with the office. He would rattle off obscure state and federal rules and laws about the financial power that the comptroller could wield. He could dissect the issues with the New York City public pension investments—the primary purview of the comptroller. Most politicians cower when they hear the words "Goldman Sachs" or "J. P.

Morgan"—the fear of rich people in politics is real. Eliot salivated at the mention of them.

Unfairly, some people think of Eliot as nothing more than a press hound for going after big financial institutions and their CEOs when he was AG. *It got him on magazine covers—that was his end goal, right?* Well, not exactly. Eliot did have a genuine, almost fanatical zeal for taking on Wall Street. He saw the financial industry as completely rigged and the cause of a lot of economic ills.

Ultimately, he simply had too much baggage to win the comptroller race. But he'd run a dignified, substantive campaign, and it helped rehab his image substantially. It reminded people that he was more than a punch line: he was a flawed, but very, very serious and passionate person.

In retrospect, I was a goner the second I agreed to consult for him. When I walked into Eliot's office, I felt something I'd never felt with anyone I'd ever worked for, a sensation I hadn't experienced since I laid eyes on Jeff that first time in the classroom. He had a wry sense of humor, a deep baritone voice that boomed across the room, and the most gorgeous eyes I'd ever seen—deep set, cerulean blue, and alight with intensity. I didn't ascribe too much importance to it at first, but after a month on the campaign, the chemistry between us was undeniable.

Political campaigns bring out weird interpersonal dynamics. You could chalk it up to the long hours and close physical proximity; it's easy to fall into a bunker mentality, like you're in a foxhole in World War I, fighting off the Germans. I'd had my fair share of relationships with coworkers on various campaigns, usually short-lived (the excitement always wore off after the race ended). But it was a whole different situation when the object of my affection was the person I was working for.

Our relationship was fraught with complications from the out-set. I was twenty-four years younger than he was. His political ca-reer had ended in spectacularly embarrassing fashion, while mine was just beginning. He had three daughters, and I had two protec-tive brothers who would occasionally joke about taking lead pipes to my boyfriends' heads. But most notably, shortly after his cam-paign ended, I'd been hired to serve as a spokesperson on Bill de Blasio's all-but-sure-thing bid to be the next mayor of New York. So Eliot and I kept our relationship relatively secret, planning to work our way up to disclosing it.

It was a little weird watching de Blasio win the Democratic mayoral nomination. I'd first met him when he'd volunteered for Edwards in Keene, New Hampshire. De Blasio was then a New York City councilman who stalked our office like a kindly, bearded giraffe. His eagerness to knock on doors and make calls for John Edwards was almost off-putting. He'd jump at the walk packets no matter how frigid the weather. In the office, he'd take on call lists and give us enthusiastic thumbs-ups during his calls. It had strong *Seinfeld* vibes of Lloyd Braun selling computers without ever plugging in his phone line. You see a lot of characters on presiden-tial campaigns. I probably wouldn't have remembered him if he hadn't been six feet six and from New York City.

So that was my only personal experience with de Blasio before I interviewed to work for him a decade later. I still thought of him as a bit unserious, and I'd heard that he was difficult to work for. The meeting was not reassuring. He showed up thirty minutes late to the little Italian haunt down the block from his Park Slope townhouse. His beard was gone, thankfully. It had made him look sloppy and unkempt.

It was just *off* from the second he sat down. Over the next hour, it slowly dawned on me that the likely incoming mayor of New

York was childish, intellectually lazy, overconfident in his own abilities, and annoyingly condescending.

He sipped a big glass of Chianti that stained his teeth purple as he asked some of the weirdest interview questions of my life.

"So what brings you here, Lis?"

"I want to work for the next mayor of New York," I told him. "I've worked on elections across the country, but I consider New York my home and I want to take everything I've learned and put it into this city."

"No. *What* brings *you* here, Lis? What was the spiritual journey that brought you to *me?*"

I was at a loss for words. I was interviewing with a mayoral candidate, not the Dalai Lama. I'm not a particularly religious person. I'd read a lot of philosophy, but I knew that Friedrich Nietzsche and Immanuel Kant hadn't brought me *here*. They probably would've burned their works if they had.

"Well, Lis, let me tell you about my journey . . ." he started. I basically blacked out for the next ten minutes as he talked about everything from Fiorello La Guardia to Buddhism. I kept trying to steer the conversation away from his pseudo-intellectual, leftist, ooey-gooey mantras, but every time I brought it back to my experience working for mayors and governors, he visibly bristled: *How dare you talk about city services and potholes?*

Finally, I just broke down. "I don't have a spiritual journey to tell you about. What I have is references like Claire McCaskill or Terry McAuliffe or Ted Strickland you can call and who will tell you I will fight to the death for you." He winced at McAuliffe's name, but conceded, "I like Strickland, he *was* a *real* man of the people." I smiled politely, knowing that Ted would have had zero time for de Blasio's condescending shtick.

I'd been through tough interviews before—with Corzine, most

notably. But I'd left the interview with Corzine knowing that even if he wasn't the smoothest politician, he was at least smart. De Blasio reminded me of the gross, unshowered guy in college who showed up to Philosophy 101 and hogged ten minutes of class time to yell about the necessity of seizing the means of production because he'd read one line of a *Communism for Dummies* book.

After we parted ways, I was shaken. I called my mom and a couple of friends on the cab ride back to Manhattan. I told them the truth: I had completely blown the interview, and I was terrified that someone like de Blasio could be tasked with running New York City in the middle of a crisis. "This guy *can't* handle a 9/11."

Still, within a week, I got offered the job. No one could chalk it up to personal chemistry, but again, it was an example of showing strength. They would pay me $10K a month on top of the other political contracts I had. The days of crying over my savings and checking accounts having a balance of -$995 were over. I tried to justify it to myself: "Maybe I can change him!" Clearly that was a big theme in my life at the time. But the reality was that my ambition was getting the best of me. I wanted to be the next New York City Hall press secretary and didn't care if I respected de Blasio or not. Likewise, I buried my concerns about my relationship with Eliot becoming public.

It was not the most rewarding of gigs. I was horrified by de Blasio's behavior that I saw behind closed doors. He reamed out staffers for daring to speak within his earshot, humiliated his campaign body person for a small error in a briefing, and made a scene when he returned to his car from a press conference and found that his preferred coffee shop order—a double shot of espresso—was not piping hot: "We have an espresso situation," he declared.

He needed his clips read to him every morning by a young

staffer because, apparently, the task of reading them himself was too taxing. He seemed obsessively paranoid; one morning as I sat in the seat behind him in the car with our NYPD detail, I received an email from him. "Watch what you say. This is not a secure space." He had a thing about his height, rounding it down from six feet six to six feet five, but still, he demanded that any podium at an event was tailored to his needs—no matter how humiliating it was to shorter speakers, including then-mayor Mike Bloomberg, another prideful guy who was experienced at lying about his height.

De Blasio was incapable of making decisions and agonized over the most basic questions from the press. He was routinely and obscenely late. Many Sunday mornings, I'd wait outside his home in Park Slope for over an hour past our planned departure time. All that waiting time meant that he was blowing off church services and community meetings, all because he "wasn't a morning person" and couldn't keep to a basic schedule. On the day of the election, we scuttled a visit to a home for domestic violence victims because de Blasio had "heavy things" on his mind—as if the women expecting his visit didn't.

None of that mattered; few voters were personally subjected to his behavior. De Blasio won the race by a wide margin, and I was offered the job of chief spokesperson for the mayoral transition. My end goals—being able to stay in New York with Eliot and serve as the New York City mayor's press secretary—were in sight.

Then, in an instant, everything changed.

It came in the form of a text from Hari: **Call me. The New York Post is looking into you and Eliot.**

Well, fuck. This isn't going as planned.

As soon as I got Hari on the line, he filled me in. A *New York Post* photographer had been surreptitiously posted outside my SoHo walkup apartment. The photographer had captured Eliot and me exchanging "loving looks" as we'd returned from a late-night dinner at a restaurant on my block. He photographed Eliot leaving my apartment early the next morning in a skullcap and hooded sweatshirt. Not exactly the wardrobe of an innocent man. I felt violated knowing the *Post* had been following me. That for *days* when I entered and exited my apartment, there was a photographer hunkered down in a parked car outside with the sole job of documenting my every movement.

Still, I wasn't exactly a PR neophyte. I knew from the beginning that our relationship would be a big tabloid story when it became public. But I compartmentalized this knowledge like other things in my life. *Okay, so I'm working for tabloid target #1 in Bill de Blasio. Maybe they won't focus on the former reigning champion, for once.* Other people could blame it on naivete, I can't. I had that early rush of love on the brain—I wasn't thinking clearly. I was also reckless. I was making the bet that I could run out the clock on disclosing our relationship until I got hired officially.

And then what?

That's as far as I'd gotten in my head. Like, on some day in February 2014 as de Blasio was dealing with a crisis at Riker's, did I think I was going to walk into his office and tell him that I was dating Eliot Spitzer? Did I think he'd sit on the floor with me and sing "Kumbaya" and tell me it would be all okay? (For the record, I can totally envision him sitting on the floor singing "Kumbaya"—just not for me.) As hardened a professional as I'd become, I could still spin a fairy tale in my head—*This is love. True love. It will prevail.* Of course, I'd blocked out the part of the fairy tale where the spinning wheel is cursed, and if you touch it, you're relegated to one hundred years of sleep.

As soon as I heard about the *Post* story, I knew I had to give the de Blasio people a heads-up about what was coming. I feigned calm even as every part of me was in turmoil—my brain was churning likely tabloid headlines and my stomach was doing somersaults.

"Hey, can I talk to you?" I asked de Blasio's soon-to-be chief of staff. "Somewhere private?"

Anyone who has worked on any government transition knows that there is no such thing as a private place in a transition office. Whether in city or federal government, it's like those spaces are designed as torture chambers. No incumbent mayor or president wants to be *transitioned* out of office—and certainly not Mike Bloomberg, who had gotten the New York City Council to overturn New York's well-established term limits so that he could run for a third term.

We found a free conference room—a drab, glass-walled cubicle with a door, really—in a corner of the office. When I told him about the coming *Post* story, his eyes widened, but he was surprisingly cool about it. He said it would be a bit of a headache but not to worry—"It's not like the de Blasios had a conventional love story. I doubt they'll be judgmental of yours." He told me to take the next day off and to work remotely. I breathed a sigh of relief— maybe things weren't as bad as I thought.

Even though I was still freaking out on a personal level, I was able to wall off my emotions and tap into my PR sensibilities and the compartmentalization skills I'd learned on Claire's campaign. I got to work. The *Post* had invaded our privacy, and they hadn't even called us yet for comment. I called Eliot—"Let's give this to the *Daily News* instead. On our terms. Are you okay with that?" He was. So from that same conference room, I called up an editor of the *Daily News*: "Okay, we'll play ball on this." It felt a little bit like writing my own obituary as I dictated the terms: they had to say that Eliot and I were "dating" and not engaged in an "affair" (factual statement). I also asked that they refrain from using the typical tabloid verbiage about women in my situation—no "floozy," no "paramour." And if they'd abide by that, I'd give them what they needed: the sourcing they needed to run with the story. Their "two sources familiar with the matter" were Eliot and me.

Dealing with the press is a high-wire act. As I touched upon earlier, it mostly depends on trust—that the source and the reporter will each keep their end of their bargain. If you agree to explicit terms, those terms must be respected. It also rests on an understanding of *how* the media works. News organizations don't work in tandem like allies during World War II. They compete against each other like the Patriots and Giants in the 2008 and 2012 Super Bowls. They want to *win*. They want to be *first*, even if *first* means they beat a competitor by a few minutes. I'd learned in Virginia that the best way to get the *Washington Post* interested in a story was to tell them the *New York Times* is already on it.

The next several hours were a blur. My boss at the time, Phil—a lovable, burly Wisconsinite who had become a personal friend—was tasked with getting me off the premises as fast as possible. He didn't know why exactly, but he could see that I was under extreme duress. It was early afternoon, just nine days before the incoming mayor of New York was to be sworn in, but he suggested we go to the bar Maxwell's around the corner to talk it out. The divey Irish bar was pretty empty when we walked in, which made its signature stench of urine and stale beer all the more pronounced.

"So what the hell is going on here, Lis?" he asked as our first round of drinks arrived. I took a few gulps from my bottle of Bud Light and started to form the words to tell him, but the second I opened my mouth, I just couldn't quite do it.

I'd spent years of my life trying to be this totally unaffected political animal. I'd been an intern, a brand builder, an attack dog. I'd talked candidates through much worse and more embarrassing situations than my relationship with Eliot. I'd learned how to tuck away inconvenient emotions, personal and professional loss, crises, general pain—I had a trash bin in my brain for all of it, and I thought I could empty it whenever I wanted and move on.

I broke as I started to confide in Phil. I had no control over the tears streaming down my face, even as I kept insisting that I was okay. It was beyond humiliating and awkward for each of us, and probably for the bartender who was warily keeping his distance.

"Yeah, we'll take another round," Phil said as the waterworks started, gesticulating dramatically for the bartender.

I exhaled deeply. I'd had to tell my parents, my siblings, and a couple close friends about the romantic relationship I'd fallen into. And it wasn't ever easy. Trust me—no dad or brother ever wants to hear the words "I'm in love with Eliot Spitzer." I thought I'd mentally prepared myself for the day when it would become public. I really hadn't. In my mind, I'd always sort of been standing near the edge of a cliff. There remained the option to walk it back—to return to the trailhead, get in the car, call it a day, and pretend it never happened.

Then there was what I did next: jumped off the cliff. "So the *Post* is about to report that I'm dating Eliot Spitzer. And I am."

Phil snorted: "Fuck. I thought you had cancer or something." He motioned to the bartender again—"We'll take two shots of Jameson on top of that round"—before turning back to me and clearing his throat. "Jesus, you always keep it interesting."

Another round of beers and shots. He asked the bartender to put the Cincinnati Bengals game on the big TV behind the bar. (I'd adopted the Bengals as my football team during my stint in Ohio. They were one of the worst teams in the league at the time—finishing the season with a 4–12 record. They were completely dysfunctional and known for their drama, both on and off the field. But they had a scrappy, underdog mentality that I found irresistible.)

For the next hour, we just sat there, watching the Bengals game. He talked shit about them; I talked shit about his team, the Green Bay Packers. We laughed about normal workplace bullshit. Almost the entire time, tears were running down my face. It was just a physical reaction that I couldn't stop.

Even in the moment, it seemed *unfair*. How many campaigns had I worked on where I'd seen really icky, inappropriate behavior— too much alcohol flowing at parties, where the lines between the most senior male staffers and the most green interns were obliterated and consent was a suggestion, not an obligation; rumors emailed around about younger female staffers' preferred taste in underwear; meetings where men would rank the attractiveness of female staffers or reporters on a scale of 1–10. HR complaints related to those incidents were buried somewhere with Jimmy Hoffa's body. The general attitude of campaigns was "Whatever, shit happens." It turns out that didn't extend to me.

Fair or unfair, I knew I was in a pickle about where to go physically. If the *Post* had had someone staked outside my apartment for God knew how long, they'd still be there. Not only that, they'd be at Eliot's place and every publicly available address where I could be found. Sure enough, within hours, multiple cars and photographers were perched outside my parents' home in Bronxville.

So I called one of my best friends: my former college roommate Nina, a lawyer who lived on the Upper West Side. When I explained the situation to her, she didn't miss a beat: "Come up, my apartment is your apartment."

Nina and I had met at the library our freshman year at Dartmouth. It wasn't some meet-cute situation where we knocked heads checking out books; it was close to midnight and we were both about to pull all-nighters writing a class paper (her) or catching up on weeks' worth of reading (me). We sat at the library cafeteria table next to each other through the night, the sunrise, and the rush of the morning crowd. We took occasional breaks to go outside and suck down cigarettes in the freezing New Hampshire air and engage in the sort of excited chatter recognizable to all freshman girls in college: "OMG—do you know [this guy]? Were you at [this show] two years ago? How awful is [this professor]?"

Nina was a city mouse. I was a suburban girl who would go to Manhattan on the weekends when I was feeling adventurous—when I wanted to get my belly button pierced, or to shop at vintage stores, or to hit up a spot that was lax on IDs on the rare occasions when my parents were out of town.

She was disarmingly extroverted. I was wary of meeting new people.

All-nighters aside, Nina was as risk averse as I was risk prone. She interned at white-shoe law firms, went to law school right after college, and lived uptown near her older brother and just across the park from where she grew up on the Upper East Side. She was in regular touch with her childhood friends and was starting to hit her stride as a litigator.

She was one of a handful of people whom I loved and trusted and considered a ride-or-die friend. She was the first person I confided in about my relationship with Eliot and one of the first people I introduced him to. Like a true friend, she never sugarcoated anything with me. She had misgivings about Eliot—"Dangerous" was the word she used. In her forthright yet gentle tone, she'd expressed her concerns: "I just think that you've spent so much of your life, like, living in these random places, and you fought tooth and nail to get to where you are. Is he worth it?"

All of that was moot when I showed up to her apartment. She had a bottle of wine open, which we drank as we sat on her couch endlessly refreshing the *Daily News* and *Post* sites. I knew what was coming, so I texted my siblings and parents not to pick up unknown numbers. By this time, most of them had met Eliot; the more awkward step was texting my recent ex-boyfriends the same warning, knowing that the tabloids might reach out to them.

The *Daily News* was first out of the gate, and while it wasn't an ideal story for me, I could live with it. My mood changed when the *Post* story popped. It was as horrible as I'd anticipated and seeing the photos of Eliot and me taken in what we'd thought were private moments was even more disturbing, knowing they were on display to anyone who cared to look. And thus began what Nina described as my "Jekyll and Hyde" phase. In the coming days, I toggled between two modes. The first, a coolly rational PR expert who could assess my own quagmire with clinical detachment. The other, a despondent and at times hysterical young woman who believed her life was ending.

After a couple hours of sleep, I woke at five A.M. to venture out to my friend's corner bodega to pick up physical copies of the tabloids. It was surreal seeing my face staring back at me from their front pages.

The *New York Post* headline writers went with "Eliot and de-Babe"; the *Daily News* settled on "Sex and the City Hall." As I coughed up the money to pay for them, I was thankful that my identity escaped the notice of the sleep-deprived guy working behind the cash register.

I'd crossed the Rubicon from being a person who "handled" stories to being the person featured in the story. My siblings would hear about it at work, my college and high school acquaintances would gossip about it on group texts and email, and my parents' friends would call them about it to share their concern. I envisioned my high school teachers spitting out their morning coffee when they read the headlines. The list went on and on in my head.

But at least I had a support system, right? I had this great boyfriend. I'd been loyal to de Blasio. I was good at my job. I hadn't done anything wrong. I'd be fine.

I'd made some faulty assumptions, but the first was that my loyalty to de Blasio or my competence at my job mattered. Behind the scenes, I'd get "keep your head up" emails from senior de Blasio aides, but when it came to the press, it seemed their attitude was more along the lines of "off with her head." I couldn't talk to the media, and they didn't lift a finger to defend me. I was called "not your ordinary bimbo" in the *Post*. Almost every write-up and TV report commented on my physical appearance—"Leggy," "often seen in high heels and pencil skirts." A story claimed that I'd been "scheming since high school to dominate the halls of power." The proof? That I'd quoted *Macbeth* and *The Fountainhead* in my senior yearbook. *Lock her up!*

Every scandal needs a narrative with archetypal characters, and the role I'd been cast in was that of the conniving whore.

It was a sort of bizarre, out-of-body experience. For years, I'd analyzed press stories about other people. Now I was analyzing them about myself. On one level it was good for me to realize *Okay, this person they're writing about isn't actually me.* But on another level, it was deeply damaging. As I've mentioned earlier, I'd had to develop a persona to fit my professional image—my own form of armor. But there's only so much protection a persona affords you. I remember after one long day on the Obama campaign, I returned to my apartment, got in bed, turned off my phone and the lights, and just lay staring at the ceiling for an hour. It wasn't bedtime; I wasn't sleepy. I just needed the noise to stop. I needed to quash the expectation that I could just talk and talk and talk and spin and be "on." For an hour, I never closed my eyes. I just found some level of peace and reenergized. Of course, no one at work knew that side of me.

I tried to keep a low profile even as I devoured every word of the stories, but it wasn't always easy. One evening, I tussled on Twitter with the infamous dirty trickster and all-around vile human being Roger Stone, whose profane tweets broke my resolve. He's notoriously vain, so I made fun of his advanced age—not my wittiest of comebacks. Our back and forth made the *New York Post* the next day. It was a good lesson in restraint. While there's an obvious value in engaging with reputable opponents and their campaigns, there is nothing to gain in getting down in the mud with a stone-cold sociopath like Stone. *Don't wrestle with pigs.*

EIGHT

In the Eye of the Storm

Eliot and I spent Christmas Day at my parents' home in Bronxville. A half dozen reporters and photographers surrounded the house, where they'd been staked out for days. Inside, the atmosphere was markedly different from our normal, freewheeling, alcohol-fueled family holidays. It felt like we were all being suffocated by the elephant in (and outside) the room. The house, unfortunately, had floor-to-ceiling windows in both the living and dining rooms. So, whether we were opening presents or enjoying Christmas dinner, the photographers were always in view. Sure, they were one hundred feet away on the street, but it felt like they had crashed our party.

My family was as supportive as they could've been that day, but they snapped the next morning when they saw photos on the front page of both of the tabloids. My siblings and parents are very private people—they'd all managed to keep low profiles and had never had any interactions with the press. It was a total culture shock for them—tacky, tawdry, and invasive. And whose fault was it? Mine, of course. One by one they called or texted me to tell me what a shitty person I was for bringing this on them. If I'd thought things couldn't have gotten worse, I was wrong.

It was the final straw for me emotionally. I was used to the ups and downs of political campaigns, the thrill of winning elections and dominating news cycles, as well as the devastation that would hit after a brutal loss. I was prepared for those moments—I'd be lying if I said that the drama wasn't intoxicating, in fact. But through it all, I always assumed that I'd have the warm comfort blanket of my family to fall back on. Thankfully, the tabloid-induced strife would be short-lived.

The next few days were dark. I stopped eating and sleeping. My brain couldn't shut down at night. It brought me back to my time living with Jeff and not being able to understand why he couldn't sleep. His brain was humming, too, but it was with a sense of positivity and ambition. I remember thinking at the time that it was all in his head—which, well, it turns out it all is—but that anyone could sleep if they wanted to. His insomnia seemed almost like a selfish act—something that transformed him into a zombie and robbed me of the vivacious man I loved. But there I was, seven years later, watching the clock tick. One A.M. Two A.M. Three A.M. Four A.M. Five A.M. Sunrise. Unlike with Jeff, there was no manic optimism behind my sleep issues, only a sense of impending doom.

The insomnia was crippling. I was exhausted but unable to sleep. I tried all the tricks—counted sheep, ran through state capitals, read boring and great books—but nothing worked. I began to experience debilitating panic attacks—episodes where I'd alternate from feeling like I couldn't breathe to dry heaving uncontrollably. No matter the hour, I was incapable of blocking out the negative thoughts that flooded my brain. It felt like every nerve ending in my body was on fire. I finally met with a family doctor who had last seen me as a healthy nineteen-year-old college student. He was horrified at my physical and emotional condition and prescribed Klonopin, an antianxiety medication, to afford my body and mind a rest. It helped to temporarily take the edge off my worst symptoms but did nothing to address the underlying causes.

I made some stupid decisions during that period. When the news of my relationship broke, I blew off the advice of smart political and media people to hire or at least fully empower someone to work with the press on my behalf. I'd always been in control of the media narrative and was reluctant to relinquish it. By the time I did, it was too late.

There is a predictable rhythm to crisis: the first wave when the story breaks and other outlets rush to duplicate it—in this case, "Eliot Spitzer shacking up with de Blasio aide."

It's in the immediate aftermath of the crisis breaking that you can either—ideally—contain the damage or—less ideally—lose complete control of the narrative. In my case, the latter occurred. Because I didn't have anyone working the stories on my behalf, trying to get a positive narrative out there about me, the tabloids ran with the most salacious ones they could conjure up ("Spitzer's gal pal known for tangling up with power pols"). So by the time we got to the third wave of stories—the stories and opinion pieces where the narrative was beginning to harden—the dominant takeaway was that I was a distraction to de Blasio and that he should not keep me on his staff ("Give Her the Heave-Ho, De Blasio").

Think about every tabloid "scandal" today, whether it involves a politician, CEO, celebrity, professional athlete, or influencer. They're either consumed by their scandal, as I was, or—after the dust clears over the rubble of the first story—the public is fed sympathetic anecdotes from "insiders" and "sources close to" the subject who combat and sometimes entirely rewrite the narrative. I *knew* that; I'd handled crises before, and I would have told anyone in my situation to take a step back and let more dispassionate professionals assume the reins.

It doesn't matter how smart you think you are, or how good you are at your job: you simply can't handle your own crisis. It's an inherent conflict of interest. Unless you're a total psychopath, it's impossible to remove your personal emotions from the equation and act rationally when negative and deeply personal things— especially false things—are being written about you.

I'm hardly the first casualty of personal hubris, but it doesn't provide me much consolation. Instead of having people actively shape my narrative, I found myself a passive observer to my trial in the court of public opinion. I let unanswered anonymous leak after unanswered anonymous leak be printed. I lost the media war that I'd thought I commanded.

Two nights before de Blasio was sworn in as mayor, the ax finally fell. The incoming mayor with whom I'd spent countless days and hours deputized one of his top advisers to call and deliver the news that my offer to work for him was rescinded. The adviser consoled me as I dissolved into predictable tears. "Trust me, Lis," he assured me. "One day you'll see this as a good thing. You don't actually want to work for this guy." He was right, in retrospect. I learned an important lesson in the hardest way possible: nothing, not even burning ambition, could justify working for a politician with no integrity.

It was nine P.M. on December 29. We were at Eliot's apartment on the Upper East Side, about to order food for dinner. By the time I hung up, Eliot had come up with an idea: "Let's go somewhere warm for New Year's." Within a couple of hours, we found a resort in Jamaica with an available room. We packed our bags and flew down to Kingston the next morning.

I slept the entire flight, my head on Eliot's shoulder. I was exhausted and spent. But when we landed in Jamaica, it seemed worlds away from New York City and the tabloids. There were no photographers. We could let down our guards a bit and relax.

On New Year's Eve, the de Blasio people started pressing me for a statement about my exit. The talking points they intended to use with the press were—as I might have expected—highly annoying. One line particularly incensed me: "Lis's departure from the de Blasio team is absolutely unrelated to recent news reports about her personal life, which is private."

It was total bullshit. How could they claim my dating life had nothing to do with the decision when it was the sole reason I was leaving? I didn't go to bed on December 21 and wake up on December 22 less qualified. Yeah, I guess I'd fallen in love with a slightly problematic guy, but if that were disqualifying, I'm not sure how anyone would get hired for any job. Also, it wasn't like I was dating Ted Bundy—Eliot was the former governor and attorney general of New York State.

I also knew how disingenuous this was coming from de Blasio. A year earlier, a top de Blasio adviser—the incoming head of New York City's Department of Investigations—had emailed Eliot, crowing about how much de Blasio had enjoyed doing his short-lived cable TV show, and telling him that "Bill and I were both wondering whether you would be open to getting involved; needless to say, you could be an enormous help." A month later, he followed up in another email pleading again for Eliot to connect with de Blasio and emphasizing how much they'd love his assistance.

In retrospect, I can see why de Blasio was pissed. Both of us had tried to get in bed with Eliot, but only one of us had been successful. And it sure wasn't the incoming mayor of New York.

It was tempting to fight fire with fire and leak the emails—they made de Blasio look craven and hypocritical. But there's also a thin line between righteous vengeance and self-defeating bitterness. Turns out that thin line overlaps *perfectly* with writing a book several years after the fact.

My short but very eventful first rodeo in New York City politics was over. Like my temper, it burned hot and fast, leaving a lasting impression on all who witnessed it. I had to move on with my life. Reading about scandals in the *National Enquirer* when I was growing up was a lot different from actually living one. The physical and emotional whiplash was unsustainable. So I drafted an anodyne statement for de Blasio to release, tweeted about how I was "looking forward" to seeing what he'd do as mayor, and tried to enjoy what time I had left on the beach.

It was a nice, much-needed break. *Peace. Finally. Peace enough, at last.*

Sure, when Eliot and I got back to the city, we saw those grainy, BlackBerry photos that someone had leaked to the *Daily News* of us shaking out our towels on the beach. But it seemed like things were quieting down.

It took all of two days to recognize that it was simply the calm before the storm. I was on the Amtrak heading to Washington when Eliot's PR person called me. She sounded alarmed: "So, the *New York Post* has a tip that they're inclined to run. They're saying that Eliot was spotted sucking your toes in a hot tub down in Jamaica. Also, you were topless."

Imagine being presented with a scenario like that, where—unless you had actually been topless in a hot tub and getting your toes sucked, much respect to anyone who's done that—it sounds too absurd to think anyone could believe it. Neither Eliot nor I was gonna win a gold medal in the "not reckless" competition at the Olympics. But *seriously*. We'd gone down to Jamaica to get away from tabloid drama, not actively court it. Our resort was more *Leave*

It to Beaver than *Debbie Does Dallas*. Our most scandalous transgression was showing up three minutes late to the breakfast buffet and sweet talking our way into cheese and jalapeño omelets.

But I'd learned my lesson. I wasn't going to rely on a yes-woman PR person, and I certainly wasn't going to be overconfident in my own abilities to contain stories about myself. So I reached out to Matt Hiltzik, the top crisis communications fixer in New York City, and asked for his help. I filled him in on the details of the coming *Post* story. Hiltzik responded with the sort of calm, knowing tone that can be cultivated only from years of actually dealing with people who'd been topless and getting their toes sucked in a hot tub: "Okay. Okay. So, it's not true, right? It's okay if it's true. We just gotta deal with the truth here." The number-one rule of crisis communications is that you must always tell the truth. (I'll get back to that one later.)

My savior. But then I got a Twitter alert on my phone: "hot tub?" Roger Stone tweeted at me. The *Post* hadn't filed a story—my phone was still warm from my call with Hiltzik. The cosmic tumblers clicked into place. Stone *had* to have been the source for the "story"—or at least involved in some capacity. He couldn't just let the story land: he had to tweet about it first. (It was a bad habit that would come back to bite him in the ass. In August 2016, he tweeted: "it will soon [be] the Podesta's time in the barrel"—a full month and a half before WikiLeaks started doing daily drops of Hillary Clinton's then–campaign chairman John Podesta's hacked emails. Three years later, Stone was convicted on all seven of the federal charges that he faced for lying to Congress, obstructing the investigation into the Trump campaign's ties with Russia, and tampering with witnesses. A key piece of evidence in his trial? The tweet where he tipped his hand about his involvement in the WikiLeaks scheme.)

Within an hour, Hiltzik had me on the phone with the *Daily News*. Once again, we were playing New York's infamous tabloids off each other, and—once again—the *Daily News* was first out of the gate:

> *After returning from a five-day vacation in Jamaica on Sunday, Spitzer and former Bill de Blasio aide Smith came home to multiple calls from right-wing reporters who'd supposedly heard the pair were engaged in improper activity at a hotel hot tub where families were present. Anti-Spitzer antagonist Roger Stone even tweeted, "@Lis_Smith hot tub?"*
>
> *"The couple both understand the scrutiny they're under," says a friend of the pair. "That's why they're even more careful about their behavior. It makes [the rumor] even more ridiculous."*

It was little consolation when I saw my photo on the wood—front page in layman's terms—of the *Post* the next day, accompanied by the oh-so-subtle headline, "Eliot and Lis' Raunchy Hot Tub Romp." The article made for disorienting reading. There had been plenty of fictitious details about me in previous tabloid stories, but this one was false from beginning to end. Upon a second and third read, I noticed the story had a couple fatal flaws.

The *Post* reported that Eliot and I had been "spied frolicking near the family pool at around 4 p.m. Sunday at the Half Moon family resort in Montego Bay, Jamaica." I knew that Eliot and I had checked out earlier that afternoon, but more importantly, I knew I had a rock-solid alibi. There was absolutely no way that I was in a hot tub at four P.M. that day. Why? Because it was Wild Card weekend. And first up that day was a showdown between the Cincinnati Bengals and the Los Angeles Chargers. Anyone with a glancing knowledge of my social media would know that I'd be more likely to miss my own wedding, divorce negotiations, and funeral than a Bengals playoff game.

I pulled the tweets I'd sent during the Bengals' loss and sent them to the *Daily News*. I was just getting started. I also produced the receipt from when Eliot and I had checked out of the hotel, along with the tab from the bar where we'd been watching the game.

When the *Daily News* went live with their story, I felt vindicated:

> *A receipt shows Spitzer and Smith checked out Sunday at 12:58 p.m.—three hours before the alleged sexcapade. At 4:19 p.m. Sunday, Smith tweeted from a nearby restaurant called Robbie's Kitchen, where she and Spitzer had watched her favorite team, the Cincinnati Bengals, lose a playoff game. That's not an easy thing to do from a hot tub while someone's licking your feet.*

The *Daily News* also featured quotes from the bartender who'd put the Bengals game on for us and hotel employees who had served us and heard nothing about any inappropriate behavior. When the story posted, my dad sent around a group family text: "I'm proud of you, Lis." One of my brothers responded drily: "Why? Because she wasn't actually getting her toes sucked in a hot tub at a family resort?" At the very least, my family still retained its dark sense of humor.

As quickly as the story had unfolded the day prior, the mockery of the *Post* for their faux exposé unfolded faster. Wonkette wrote about the *Post*'s bungled reporting, noting my questionable taste in sports teams: "We guess the larger point is that the Cincinnati Bengals have fans outside of Cincinnati." I'm probably the only person in history whose alibi was watching a Bengals game. In the seven-plus years since that incident, I can count on three fingers the number of Bengals games I've missed. The unluckiest team in the NFL was my lucky charm when I needed one most. I'll always tune in to watch them, no matter how bad their season is, because *you just never know.*

I entertained taking up legal action against the *Post* and was talked out of it quickly by my brother-in-law. Whatever truth I had on my side was irrelevant—it would be ruinously expensive if I took it to court. And really? Did Eliot and I want to go through discovery? He wasn't a prude, obviously, and nor was I. I could see how this would play out and held my horses.

Anyway, it was largely irrelevant at that point. Two days after news of my fictional "hot tub romp" imbroglio unfolded and was promptly debunked, I was knocked off the front pages by a far bigger story—one I couldn't have scripted better if I'd tried: Chris Christie and Bridgegate.

Even I had to laugh when I saw *Vanity Fair*'s post: FIVE PEOPLE WHO COULD NOT BE HAPPIER ABOUT BRIDGE-GATE. I was the first person mentioned, above then-president Obama, Roger Ailes, Target, and Fort Lee mayor Mark Sokolich. The karmic irony was not lost on me. Christie had unwittingly become my savior. God, indeed, has a sick and twisted sense of humor.

I agreed to only one interview about the whole contretemps. I spoke to the *Times* writer Ginia Bellafante for a column to give my side of the story. We sat down for lunch in a booth at the Mercer Kitchen in SoHo for the interview, and before we even ordered, I burst into tears. For a brief moment, the wall between the reporter and subject came down as Ginia placed her hand on mine and asked if I was seeing a therapist to help me get through the ordeal. I was, I told her, and even after she flipped back into reporter mode, turned on the tape recorder, and started asking uncomfortable questions, I remember being so thankful for the sensitivity and humanity she extended toward me. The column she published was extremely sympathetic and closed on a high note with a quote from me: "I want to show people that who I date has no bearing on my professional capacity and I want to use this as motivation to become one of the best in the business. I want to prove them all wrong."

I meant every word of that statement, even if it would take a few years to come about. But the reality was that I was in a world of pain. I was completely unprepared emotionally for the Twilight Zone that I'd entered. I became extremely attuned to—and paranoid about—any camera flashes in eyesight. I had a sixth sense for when people were turning their phones on Eliot and me.

At restaurants, I would frequently ask maître-d's to go up to the offending party and make them delete whatever photos they'd taken. If they didn't, I would approach them myself. I confronted anyone on the street or subway who tried to snap stealthy shots. Eliot almost always played the good cop to my bad cop. He was used to being a subject of public scrutiny and more tolerant of it. Of course, there were a couple exceptions during our time together, but his restraint was largely notable. Mine was nonexistent.

In spite of being humiliated by de Blasio, I had my other work to fall back on. I was still consulting for both the DGA and O'Malley. A few days after the *Post* toe-sucking story broke, I went into CNN with O'Malley for a Sunday morning hit on *State of the Union*.

The greenroom was the usual hodgepodge of political talking heads and staff. Sean Spicer, then the communications director for the Republican National Committee (and later Donald Trump's first White House press secretary), was in the room, readying for his spot on the show. I zeroed in on him. Right-wing blogs and Twitter personalities had been going nuts with the *Post* story, but I'd noticed mainstream Republicans like him hadn't been. I thanked him for that, and he was clearly a bit taken aback: "Lis, there's honor among thieves here. What de Blasio and the *Post* did to you was fucked up. You have my word that we will never touch this."

During O'Malley's hit, I was checked on by both a top programming person at CNN and a female columnist who was also on set that day. Their intentions were completely pure, but I felt myself chewing the inside of my cheek till it bled as they asked, "Are you okay?" and told me, "Hang in there."

That was my professional reality at that time. And as uncomfortable as I felt in New York, it was far worse in Washington, which is much more staid and conservative. I could feel the stares and hear the whispers when I walked into rooms. It was like I was a modern-day Hester Prynne.

My big coming-out in DC took place in February, when I was organizing the press strategy for Democratic governors around the biannual National Governors Association meeting, including a sitdown with President Obama at the White House. The DGA was kicking off the weekend with a luxurious lunch held in a spacious, light-filled event space on the top floor of the Newseum. As the Democratic governors and their staff started to file into the room, I turned to Danny Kanner, my old Obama coworker. "Help me. Don't leave me or make me talk to anyone," I told him.

Everything was going according to plan. That was, until Terry walked into the room. Terry was riding high off his November win and swearing in as the seventy-second governor of Virginia. People lined up to hand him business cards and kiss his ring. He did a couple obligatory hellos and handshakes, but as soon as he spotted me, I heard his voice boom across the room:

"LOOK WHO IT IS! LIZZY? IS THAT YOU?"

Every eye in the room traced his movements as he walked toward me and continued the Terry McAuliffe show.

"LIZZY IS FAMOUS! SHE'S DATING THE FORMER GOV-

ERNOR OF NEW YORK! SHE'S BEEN ON THE COVER OF
ALL THE TABLOIDS! STEP ASIDE, WE HAVE A CELEBRITY
IN OUR MIDST!"

Whatever hope I had of fading into the background was shot.
There were a lot of awkward stares. I was slightly mortified, but
also very, very amused.

He leaned in for a hug and a kiss on the cheek. His public,
larger-than-life persona disappeared as he whispered conspiratorially: "Lizzy, someone needed to lance that boil for you. Had to
happen. Now everyone can relax. Plus, you looked DYNAMITE in
those photos."

The easiest, most natural and normal response in that scenario
would have been for Terry to wave at me from afar, leaving me
to hide in my corner. But he'd called soon after the tabloid stories broke and knew how hard I was taking them. He'd heard all
the negative, gossipy whispers in DC. He also knew—better than
anyone else—how to wrestle the alligator in the room, and that
people would take note of what he did and said.

As he walked away to work the rest of the crowd, I felt a sense
of calm wash over me that I hadn't felt since the news of my relationship with Eliot had dropped. I giggled with the first real joy
I'd felt in months.

The reality is that I'd allowed the tabloid coverage and the de
Blasio firing to rattle my confidence. I know that the people who
went out of their way to express their sympathy were well meaning, but they never made me feel better—if anything, they made
me feel worse. I didn't want to feel like a victim, I wanted to have
some fun again. Terry, always the minister of fun, gave me a much-
needed moment of levity and perspective.

Had I been excoriated and humiliated by the tabloids? *Yes.* But
did any of that change who I was? *No.* I was good at my job, and I
had never fit the mold of a typical political operative anyway. Yes,

the shit-talking had increased substantially of late, but it wasn't like a new phenomenon. Why should I suddenly give a fuck about it now? I was still the same person.

There was and still is an unfortunate double standard in politics. The men in politics who find themselves in Page Six or who lead colorful lives are portrayed as rakish, larger-than-life characters: "Oh he cusses, drives a Porsche, and went out on a date with a supermodel? He's a modern-day god! Let him make $10 million a year!" But somehow, women in politics are expected to be totally buttoned-up, Type A nuns.

After John Kerry lost in 2004, a *Newsweek* story postmortem on his campaign focused critically on Stephanie Cutter, then his campaign communications director, noting her "dirty blond hair, short skirts, and high boots." Meanwhile, her older, drab, gray-haired male coworkers wearing ill-fitted Brooks Brothers suits got no such treatment. To be clear, I don't think that the sartorial choices of presidential campaign staff ever affect the outcome of elections, but I do think that Kerry's lackluster, losing campaign would've benefited from a few more skirts in the room—a little more sizzle, and a lot less meh.

The month before I ran into Terry, I was called in to meet with a senior-ranking member of the New York congressional delegation. It was a week after the since-debunked Jamaica "exposé" ran. *Why on earth would anyone, anyone want to talk to me right now?* At a meeting in the congressman's DC office, I self-consciously mentioned that I had "a *New York Post* problem."

"I know what the *Post* has written about you," he told me. "I don't give a shit about the *Post*." He might not have, but his staff did. I wasn't offered the job.

Now, I started to use my experiences with the tabloids as a selling point. I talked about what I learned from the ordeal, how it made me better at my job and helped me better understand the life cycle of public relations crises. Owning my "negatives" and my persona became central to my professional career. And if that persona put someone off—fuck 'em. It wasn't a match to begin with.

NINE

Good on Paper

By the summer of 2014, the 2016 "invisible primary" had begun—the monthslong process that precedes an official presidential campaign announcement. Potential candidates and their closest aides call around to donors, reporters, and political staffers to gauge whether there's enough interest in them to mount a credible effort. Under normal circumstances, dozens of Democratic governors, senators, mayors, and representatives in the House would have been taking part in the time-honored tradition. But as soon as Hillary Clinton started to make noise that she might run again, even the most ambitious of pols threw in the towel. The bond between the Clintons and the Democratic donor community was unbreakable—it would be nigh on impossible for any other candidate to get the cash they needed to fund a campaign. Even so, O'Malley had his heart set on running and was convinced that voters would be ready to turn the page on the Clintons. I shared that belief and hitched my wagon to his. It was a long-shot effort, but we'd never know if we didn't try.

He had his first national breakout moment that summer at the National Governors Association gathering in Nashville. The backdrop to the meeting was the humanitarian crisis in Central American countries like Guatemala, El Salvador, and Honduras that had led to a surge of unaccompanied young children appearing at the Mexico–US border seeking asylum. The Obama administration, no doubt fearing a right-wing backlash, announced that they would waive the normal asylum process and begin deporting the children as soon as possible. In the days leading up to the NGA meeting, news organizations published photos of kids as young as five and six years old being held in cages on the border. O'Malley was outraged; the policy stood against everything he believed in.

It was not lost on us that then–secretary of state Clinton had spoken out in favor of the deportations, saying: "We have to send a clear message. Just because your child gets across the border, that doesn't mean the child gets to stay."

Before the NGA meeting, we gave a courtesy heads-up to the White House that O'Malley was going to speak out against their policy. Top Obama officials carpet-bombed our phones with texts and calls urging him not to do it. He remained resolute and, when asked about the policy at the press conference, held nothing back: "I can only imagine, as a father of four, the heartbreak that those parents must have felt in sending their children across a desert where they can be muled and trafficked or used or killed or tortured. . . . We are not a country that should send children away and send them back to certain death."

It was powerful stuff—the most passionate and emotional I'd ever seen O'Malley. He could barely contain his rage. At times, his voice quivered with emotion. He was one of the first national

Democrats to speak out against the policy and the most high pro-file. The White House was positively livid as the headlines poured in and other Democrats rushed to side with O'Malley. Over the course of the next week, I found myself going head-to-head with the president for whom I'd once worked. The White House tried to leak to the media that O'Malley had opposed relocating Central American refugees to Maryland, but their claims didn't hold up under scrutiny. In the end, O'Malley ended up on the right side of the argument—he was praised by immigration reform groups, Latino activists and elected officials, and editorial boards for his early, courageous stance on the issue, and it helped him stake out a place in the national debate.

It's a bad sign for a presidential campaign, however, when the emotional high point comes over eighteen months before the Iowa caucuses. Unfortunately, that was the case with O'Malley. Every-thing was pretty much downhill from there.

First, there were the 2014 midterm elections. Republicans steamrolled to victory everywhere, including in Maryland, where O'Malley's lieutenant governor was running to replace him and lost to Republican businessman Larry Hogan. It was galling to O'Malley on a personal level, since he had worked so hard to fend off the 2010 Tea Party wave. His legacy and accomplishments were erased overnight. The overwhelming sentiment shifted to: "If he can't hold the governorship in Maryland, how can he win the presidency?"

That spring, as our small team was making arrangements for his presidential announcement, we got even worse news. Freddie Gray, a twenty-five-year-old Black resident of Baltimore, died in police custody. The details of his death were murky. The outcry, however, was immediate. The year prior had seen the birth of the Black Lives Matter movement after Eric Garner was killed in a police chokehold on Staten Island, New York, and Michael Brown had been shot to death in broad daylight by a police officer in Ferguson, Missouri.

Soon after the news broke of Gray's death, the people of Baltimore took to the streets to protest it. O'Malley—the former two-term governor and, more importantly in this moment, the former two-term, tough-on-crime mayor of Baltimore—was on a flight to Ireland as news of protests broke. As soon as he landed and heard what was happening, he booked the next flight home to be with his family, his city, and his state as they mourned. He loved Baltimore. We soon learned, the feeling wasn't necessarily mutual. As O'Malley visited sites across the city, he was as likely to get heckled as he was to be thanked for his efforts to help the community.

The divide was also on display in the media. Matt Bai, a Yahoo news columnist who spent a day with O'Malley in the middle of the chaos, wrote: "I came away thinking, though, that in some strange way the events of this past week had the potential to make O'Malley a more compelling [presidential] candidate, rather than less." The Daily Beast, on the other hand, went with the headline: "O'Malley 2016 Commits Suicide in Baltimore." Sure, it was the Daily Beast and not the *Washington Post*. But as I'd learned during the 2016 campaign, news trickled up as much as it did down. Once the headline hit the Drudge Report, it didn't matter where it came from. It was in the bloodstream.

We got through the Baltimore protests. A month later, O'Malley announced for president. On paper, he ticked all the boxes of a successful presidential candidate—he had the bio, the résumé, the looks, and the perfect family. In just the last couple years alone, he'd championed and enacted some of the most important, progressive priorities—marriage equality, gun control, abolition of the death penalty, a higher minimum wage, decriminalization of marijuana, a state-level DREAM Act and driver's licenses for undocumented immigrants.

Of course, what looks perfect on paper doesn't always translate in real life. A reporter profiling O'Malley for a national publication confided after a few days on the trail with him: "I guess I just wanted or expected him to be *more*." There was the O'Malley behind the scenes and off the record—warm, genuine, funny, and smart. Then there was the O'Malley who appeared the second a TV camera or recorder turned on. Guarded and stilted in his presentation, oddly formal in his language, he was like a mid-2010s candidate doing an impression of a 1990s era B-list actor doing an impersonation of 1960s-era Kennedy—he just couldn't quite connect. He was branded with the I-word, the death knell for any politician: inauthentic.

Like most everyone else in the trenches, we sort of shrugged when we saw that Bernie Sanders, the self-identified Democratic Socialist US senator from Vermont, was entering the race. Nothing about him screamed presidential—on paper at least. He was a fringe figure in national politics. He was seventy-three years old, with unkempt white hair and a wardrobe that looked like it was plucked from a Vermont Goodwill store in the 1980s. His political style could be summed up largely as grumpy and yell-y; he didn't pretend to engage in small talk with reporters or voters, and his long-winded rants against "millionaires and billionaires" recycled material he'd been spouting for decades.

This guy is a joke. He's not going anywhere. That was the mainstream political opinion. It was only natural, then, that Sanders's campaign took off like a rocket ship, leaving O'Malley in the dust as the alternative to Clinton. Within two months of announcing his presidential campaign, Sanders raised $15 million—all in small, online donations. Even in the early days, he was attracting jaw-dropping crowds—over ten thousand people to Clinton's two hundred to three hundred. And that was just the beginning. By the end of the primaries, Sanders was outraising Clinton and addressing crowds of thirty thousand.

Why him and not me? That's the question I imagine always lurked in the back of O'Malley's mind. In hindsight, the answer was obvious. Clinton was a dominant presumptive nominee: she had 100 percent name ID, the support of the majority of the political establishment, and money up the wazoo. And yet the loyalty that she inspired in some people was matched by the antipathy she stoked in others. Even in a Democratic primary, there was a clear anti-Clinton lane, but the lane was wide enough for just one candidate. If you were concocting the most anti-Clinton prototype of a candidate, you couldn't come up with much better than Bernie Sanders. He was as antiestablishment as she was establishment, as radical in his political positions as she was cautious, as seemingly independent as she appeared beholden to special interests and rich donors. Sure, he was a little weird and a little unpolished, but even that was a refreshing contrast to the Clintons and their very insulated world. Our best—and honestly only—hope with voters was that they'd be like Goldilocks in "The Three Bears," settling on the porridge that was neither too hot (Sanders) nor too cold (Clinton), but just right (O'Malley).

The die was cast. O'Malley was running as a challenger to Clinton, but in reality he was a relatively traditional politician with insider credentials—mayor of Baltimore, governor of Maryland, chair of the Democratic Governors Association. His voters and donors overlapped largely with Clinton's. Minus a few policy differences here or there, there wasn't that much daylight between them.

The institutional deck was stacked against O'Malley. The Democratic National Committee, in conjunction with the Clinton campaign, constructed a debate schedule that ensured as little viewership as possible. They set up just four debates prior to Iowa and scheduled them during big sports games and holiday weekends. The implication was clear—debates were risky and the DNC and Clinton camp wanted to protect Hillary for the general election.

The first debate was held in mid-October in Las Vegas. It was the only one with a sizable audience and O'Malley's best chance to break through. He didn't—he completely disappeared onstage. Hillary was declared the consensus winner, and Biden announced within days that he wouldn't mount a challenge to her.

We shook up prep a little for the upcoming debates. I took on the role of playing Hillary Clinton. It was a lot on my plate, but even more so when you considered what was going on in my personal life.

My relationship with Eliot was falling apart. After the rush of our whirlwind courtship, we discovered that we weren't exactly the best match in the world. Temperamentally, we were like a lit match and dynamite. Our social lives could not have been more different. And I was still in the political world—a world from which he'd been expelled. There were other issues in our relationship, but I'll leave it there.

I cut the cord right before the second debate, the night before a critical prep session. You can imagine the O'Malley team's surprise when I showed up straight from the Acela bearing not only a very large suitcase, but also Cersei, the kitten that Eliot and I had adopted from a rescue months earlier. I studiously ignored the NO ANIMALS ALLOWED sign outside the private school whose auditorium we were renting and asked a colleague to hide Cersei.

I was obviously in a bad mood; breaking up sucks. I channeled my emotions into savaging O'Malley onstage. At one point after I delivered an especially harsh line, O'Malley broke character: "Jesus Christ, Lis. She's not going to go there, is she?" (It was the first, though not the last, time that debate prep would serve as a form of therapy for me.)

Despite all the time we'd put into prep, the moment I remembered most happened off camera during the Democratic showdown in Manchester, New Hampshire, a few days before Christmas. Per usual, I was one of O'Malley's designated, on-site staffers. And naturally, I found myself a ball of nerves, chugging coffee as the debate unfolded. It wasn't a particularly winning one for O'Malley.

When the moderators announced the first commercial break, I sprinted to the only ladies' room on the floor, the locker room for the St. Anselm College's women's hockey and basketball teams. The Secret Service agents assigned to Hillary Clinton tried to block my entry to the ten-stall bathroom but were overruled by Hillary's longtime aide Huma Abedin, who recognized me and waved me in. *Thank God.*

I peed in record time and when I left the bathroom, Hillary was standing right outside the door, waiting for me to leave. I didn't get the distance act—there were a ton of stalls she could've gone into, and I wasn't exactly a security risk.

As I settled back in my seat in the greenroom, I saw the debate resume. There was one glaring issue, though—only O'Malley and Sanders were onstage. Clinton's podium was empty. After a minute or so, she finally waltzed back onto the stage and took her place, saying simply, "Sorry."

I shot my coworkers a look. "Holy shit, guys, I think I caused that," I told them.

The debate itself was immediately overshadowed by Clinton's late return to the stage. The media and Republicans had so little trust in anything that anyone named "Clinton" did or said, Twitter immediately lit up with questions. *Where had Hillary been? Why was she late? Was it* really *a bathroom break?* Republican operatives started a rumor that she had been doping up with performance-enhancing drugs behind the scenes (an accusation that Trump would lodge against Joe Biden four years later during their debates).

After all the beatings I'd taken in the press, I had no desire to be anywhere near the story. But as so often seemed to be the case for this agnostic, my prayers to be excluded from the narrative were not answered. The next morning, Annie Linskey, a presidential campaign reporter for the *Boston Globe*, called. "So I got a tip that a female O'Malley staffer was to blame for Hillary's delay. And that could only be you, Lis." I did my best to provide her with proper context—that this wasn't some nefarious political plot on my end, just a misunderstanding. When her story posted, it was immediately picked up by the Drudge Report and given prime real estate as the main headline. *So much for keeping my head down.*

Every few minutes, I saw a new Google alert for my name. I was called a "potty blocker" and referred to as the "caffeine guzzling deputy campaign manager." Naturally, the "story" got big attention in the tabloids. The coverage was so prolific that a coworker wryly quipped: "How much does it suck for O'Malley that your bathroom trip is getting more media play than his debate performance?" It got even worse when Donald Trump fixed his attention on it at one of his raucous rallies: "I know where [she went]—it's disgusting, I don't want to talk about it. No, it's too disgusting. Don't say it, it's disgusting."

I'll never know why Hillary wouldn't share that multistall bathroom with me that evening; we'd shared one before. But I suspect it wasn't because she had gastrointestinal issues. The networks that host debates are emphatic that candidates cannot interact with staff during breaks; it's seen as "cheating." My personal theory is that she wanted to skirt that rule by having a private conversation with staff in the bathroom. The only thing I know for sure is that my intentions in that moment were pure. I didn't want to torpedo her debate or make her late to the stage, I just really, really needed to pee.

O'Malley dropped out the night of the Iowa caucuses after a disappointing finish. We'd made the call to suspend the campaign the night prior, and took pride in the fact that it didn't leak. He got to announce his exit on his own terms.

I found myself in a similar position post-2008. I was unemployed, newly single, and depressed about my life prospects.

It was customary for me to take a vacation after a tough election. I knew that I needed more than the normal beach excursion this time. Warm weather and fruity cocktails weren't gonna cut it. I needed to take the mother of all getaways—I needed to escape New York and politics. I needed to take a vacation that put *Eat Pray Love* to shame.

After a couple days of research, I came up with the answer: Uganda. It's hard to think of a destination more unlike New York City than East Africa. And as it also happened, for the past twenty-five years, I'd been dying to see the mountain gorillas of Rwanda, Uganda, and the Congo. I'd read Dian Fossey's memoir, *Gorillas in the Mist*, in third grade and became fixated on her work in Rwanda. I even wrote my college admissions essay about my obsession with Fossey.

So just a few weeks later, I found myself traveling around Uganda by myself for two and a half weeks. Was it a little bit crazy? Sure, but it was just what the doctor ordered. It was the low season, so I was usually the only guest at the lodge. I befriended my tour guide and the lodge employees—most nights, we'd stay up past midnight drinking wine, smoking Ugandan cigarettes, and swapping stories. I unplugged from the tabloids, Twitter, and cable news. I read too many books to count. And, of course, I fulfilled my lifelong dream of seeing the mountain gorillas—not just once but twice. I still remember the exhilaration I felt on my first trek when I heard their low grunts in the bush. Laying eyes on them was even more incredible: they were the most beautiful and majestic creatures I'd ever seen. On that first day, I had shared the experience with three tourists from Hungary and four park rangers wielding machetes and very large guns. I was emboldened the next day to visit a different area of the park where there were multiple families of mountain gorillas that tolerated human presence. I told my guide that day that I didn't want to trek with any other tourists; I wanted to go alone with the rangers, if possible.

Before we departed the base, I received the same warning as the day prior: "When you're around the gorillas, don't overreact or make any sudden movements." Anyone who has spent more than five minutes around me would know that I'm a highly excitable, easily startled person. And yet I nodded in agreement: "Of course. Of course."

Forty-five minutes into the trek, one of the rangers raised his hand and made a shushing noise, then motioned down the hill and raised his hand to his ear—a sign to listen for the sounds of gorillas he believed to be nearby. Indeed, I could hear twigs and branches crackling just below us on the hill. At this point, I started to get a little nervous. The day before, the gorillas had been in a fairly wide-open space in the forest; this terrain was hillier and the brush was much denser. Still, I followed the rangers down the hill as two of them hacked away at the brush with their machetes.

Midway down the hill we took a sharp left turn and the lead ranger pointed up at a tree ten or so feet from us. The moment I'd been fearing arrived—the second I met the gaze of the five-hundred-plus pound blackback staring at us, I could feel my heart jump out of my chest. It brought me back to the haunted houses of my childhood and my involuntary, disproportionate reaction whenever a scarecrow or Michael Myers–esque character would jump out in the dark. I was a full-grown adult now, but before I could stop myself, I was recoiling and screaming and the gorilla was jumping out of the tree and charging toward us. Things only went downhill from there, literally.

In this situation, the rangers tell you, you should always hold your ground. The gorilla is just trying to show dominance and will not actually make physical contact with you. So, of course, I did the last thing in the world that I was supposed to do—I grabbed onto the machete-wielding lead ranger, tackling him in the process. We somersaulted down the hill, our bodies entwined, his machete glinting in the sunlight that shone between the trees, our combined momentum taking out every piece of shrubbery in our path. When we finally slowed to a stop, ten seconds later and one hundred feet downhill, his chuckles were drowned out by the riotous laughter from the rangers above us. I, on the other hand, still felt like I was about to have a heart attack—or die from embarrassment. I'd asked for this special excursion, after all. One day of visiting with gorillas, and I had convinced myself I was Dian Fossey. Not the first or last time that my cockiness would get the better of me.

The rangers ended up extending the visit with the gorillas by an extra thirty minutes to make up for my meltdown. I tensed up every time I heard so much as a branch crackle, but they couldn't have been more entertained. On our way back to the base, as I pulled leaves and burrs from my clothes, hair, and even my underwear, they took turns theatrically imitating my complete and utter freak-out. They did play-by-plays of my flailing physical gestures and wide-eyed screams. "GUYS! I'M SORRY. I'M SO SORRY!" I repeated between my shrieks of laughter. The trauma and stress of the last few years melted away. It was like my own twisted form of meditation: for hours afterward, I was completely free of all the bad thoughts and anxieties that had been keeping me up at night.

TEN

The One

I had just had the most uplifting, cleansing twenty days of my life. I wish I'd extended the trip longer. The dread set in almost immediately when I landed at JFK. *What the heck am I gonna do next?*

I'd decided two things in Uganda: I wouldn't work in politics again—or any time soon, at least—and I wouldn't date anyone or pursue another intimate relationship. I felt so damaged, there was no way that I could do either successfully or authentically.

The months from April 2016 through the end of the year were largely unremarkable. I did my best to avoid political work, tried my hand at PR for nonpolitical clients—some of the most boring work I'd ever done—and dabbled in TV punditry. Every week, I'd appear on MSNBC, Fox News, or CBS's digital network, where I got a contract before the election.

From the outside, TV work seems sexy and exciting. And there were days when I found myself in the right place at the right time—the night that the Trump *Access Hollywood* tape dropped, I was commentating on MSNBC for hours as legends like Tom Brokaw piped in via remote.

I remember that night well. I was repulsed and outraged by the *Access Hollywood* clip. It was objectively disgusting. But always there was this aspect of playing to the cameras: saying the things the audience wanted to hear most, delivering the most dramatic and over-the-top commentary, knowing that it would get the most play and maybe go viral.

Outrage was especially in demand in 2016, and I found myself regularly pandering to the lowest common denominator and making assertions I would never have made in a strategy meeting or even over dinner. I played less to the crowd than to other Democratic commentators, but I didn't feel proud of the work I was doing. Every time I had to commentate on a debate or speech, I found myself calibrating what I said to what people wanted to hear. And in 2016, what people wanted to hear was that Hillary Clinton was running a flawless campaign and there was no way Donald Trump could win. Of course, neither thing was true.

At night when I washed off the fake lashes and TV makeup and fielded texts from friends who'd seen my appearances, I couldn't shake the sense that I was missing out somehow. I wanted to be in the mix of it all, not offering filtered commentary from the sidelines.

Trump won the presidency. The campaign I'd worked on had flamed out with barely a whimper. I was considering leaving the business altogether.

"Tough name."

That was my first response when I was contacted by a colleague about a potentially promising candidate for the open position of

DNC chair: Pete Buttigieg. It was mid-December 2016, and the Democratic Party was in a state of shock (and also in the midst of an identity crisis) following Trump's win. I was sitting on my couch in pajamas, drinking instant coffee, and doing some last-minute online Christmas shopping.

As disillusioned as I'd become, I still felt the familiar pull, the singular thrill of politics. I ditched my shopping cart and headed to Google to check out this Buttigieg guy. *It couldn't hurt to take a look, right?*

The first hit was a Frank Bruni column in the *New York Times* from June 2016: "The First Gay President?" Bruni led his piece, "If you went into some laboratory to concoct a perfect Democratic candidate, you'd be hard pressed to improve on Pete Buttigieg, the 34-year-old second-term mayor of this Rust Belt city, where he grew up and now lives just two blocks from his parents."

The second was a *New Yorker* profile of Barack Obama written by David Remnick that had posted right after the November election. In it, Obama named four Democrats as the future of the party: Tim Kaine, Michael Bennet, Kamala Harris, and Pete Buttigieg. *How have I not heard of this guy?*

Granted, my knowledge of South Bend was limited. My only experience with the town was a night I'd spent at a chain hotel that I checked into after one A.M. on my way from New York to South Dakota.

I went down a rabbit hole. I read the few other national clips—and trust me, there were only a few. I migrated to YouTube to watch some of his interviews and public remarks. I read local coverage of him in the South Bend media. It wasn't lofty or heady stuff, but I was transfixed.

I texted Jeff, by then happily married and leading a nonprofit to help ex-cons reenter society: "Do you know Mayor Pete? Or people who know him. I want a read on his douche factor." To my jaded political mind, no one could be thirty-four years old, floated in the *New York Times* as the first gay president, named as one of four future voices of the Democratic Party by Barack Obama, and not be a douche.

Jeff had never heard of him. No one I knew had.

A few days later, Buttigieg published a Medium post, "A Letter from Flyover Country," laying out his thoughts on the future of the Democratic Party. Soon after, I found myself on the phone interviewing with him for over an hour. It was different from most of the conversations I'd had with politicians. He admitted at the top of the call that he was green when it came to national politics and that he didn't know everything.

I did most of the talking. He asked thoughtful, granular questions about media strategy—how to get attention in a multicandidate field where he was the least well known quantity; how the national media worked—everything from how I'd get meetings with national reporters or bookings on national TV to how I'd help craft a message for his campaign and sell it in a way that was compelling to the media and a national Democratic audience but still true to him. He had the nitty-gritty interest in tactics of a campaign manager, but a rock-solid grasp of who he was and what he had to say. He was clear-eyed about the long-shot nature of his campaign, but he wasn't interested in hiring national political consultants who would try to change who he was and what he believed to get cheap attention. He showed a maturity and self-awareness that outstripped politicians three decades his senior. In a word, he was *refreshing*. The only open question was whether he would have the charisma and public presence to get any traction.

Before the curtain fell on 2016, he'd hired me to work on his long-shot DNC chair race.

In the seven days between my hiring and his announcement, there was a lot to be done. He added a few more staffers: a scheduler, a political director—the person responsible for harnessing endorsements and managing Pete's call time with politicos—and a fundraiser. He gave courtesy calls to party leaders and former DNC chairs—people whose egos he wanted to stroke in hope of an endorsement down the line. On my end, I began working on landing the best media exclusive we could get. It may seem like an insider or insignificant consideration but landing a good "exclu" is as good as it gets in politics. Yes, the provenance of a clip matters less than ever, but Pete needed as much credibility as he could get, so it was critical to get one of the big-time papers, the *New York Times* or *Washington Post*, to run with it. Luckily, the *Times* agreed to break the news.

Everything was laid out and perfectly planned. And then my mom called me the night before Pete's announcement. My seventy-eight-year-old dad, whose gnarly cough I'd noticed on Christmas, had full-blown pneumonia and was in the hospital. He was in very rough shape; his health had been in a state of decline since he'd been diagnosed with Parkinson's disease in 2013, and my mom struggled to keep up with all his ailments and hospitalizations on her own. My siblings all worked fairly traditional jobs: my sister is a doctor at NYU Hospital, my older brother is an executive in the insurance industry, and my twin is the CFO at a health-care company. For them, spending time with my dad would mean taking time off from work. With my job, however, I really could work from any place that had Wi-Fi and cell service.

So on January 4, the day that Pete announced his candidacy, I found myself flacking and calling reporters from a chair at the foot of my dad's hospital bed. The attending nurses shot me annoyed looks as I spoke loudly and occasionally "colorfully" on the phone. They even shushed me a couple times.

But my dad was tickled getting to listen in on my conversations with reporters. In between calls, he'd ask, "So who was that one with?" When I'd tell him the name of the reporter, he'd laugh. "That's how you talk to them?" It was a welcome distraction for him. That night, Pete was on a primetime MSNBC show, which my dad and I watched on the small TV in his room. My dad was impressed—and Pete had won an important primary in my mind.

The next week, I flew out to Phoenix to prepare and staff Pete at the first DNC chair meeting. We had met in passing a few days earlier at Obama's farewell speech in Chicago, an interaction that lasted only a couple minutes. So this was the first time I'd actually get to spend time with him in person.

I headed up to his hotel suite and was greeted at the door by his campaign manager, Jennifer Holdsworth—a diminutive, and *very* loud redhead. "WELL, HELLO!!! Pete, Lis Smith has entered the building" was how she introduced me. Pete was dressed down in jeans and a long-sleeved beige T-shirt. He had impossibly thick, fluffy hair and a dark scruff on his upper lip—like he hadn't shaved in a few days. He looked more like a college kid than the next head of the Democratic Party.

What struck me immediately was how quiet and unpretentious he was. Most politicians take up massive physical or emotional space. You couldn't be in a room with Terry without knowing he was there. When Barack Obama sneaked into the room where Claire was getting sworn in as a senator, every eye went to him. He was tall and beautiful and still somewhat unaware of the command he had on mere mortals. But as soon as people started gawking at him, he jutted out his chin and squinted his eyes in that way politicians do when they want to seem engaged by anything but themselves. Less "I feel your pain" and more "I feel your stare."

That day in the hotel room, we held three different prep sessions with Pete totaling five or six hours—it's a process that we affectionately refer to as a "murder board" in politics. I had a list of thirty questions to ask him—as we'd run through them, I'd write down his best lines or things he needed to work on, and give him concise feedback. We touched on everything from his thoughts on the 2016 election, what he believed the future DNC should prioritize, and—for fun—a few wild-card questions about the economy and US-Israel relations. The latter wasn't exactly within the purview of a DNC chair, but I was curious to see what this guy was made of. Plus, I'd abused my Rolodex and set up interviews with national reporters the next day, and it was critical that he impress them.

I'd worked for national politicians with plenty of press exposure who would've pissed themselves from anxiety at the thought of five back-to-back interviews with reporters at marquee publications. Pete didn't blink an eye. He had a quiet intensity to him that I'd never seen before in a politician—a completely different type of poise. He focused squarely on everyone he spoke with—never looking at his phone or breaking eye contact—and he spoke in soft, deliberate tones. He knew his way around talking points, sure, but somehow delivered them in a completely fresh way.

When I've been asked to describe the moment I first met Pete in person, I use an analogy from my music fandom. Growing up, my favorite band was Guns N' Roses. I went to their shows, knew all the lyrics to their songs, fantasized about marrying Axl Rose. Then, one day I heard Radiohead for the first time and felt disoriented. Radiohead's look, their vibe, and their sound was so different from what I was used to, but it was *captivating*. That was how I felt the first time I listened to Pete. I'd finally experienced that moment that prominent political advisers like David Axelrod and Karl Rove talked about—I'd found *the one*.

I wasn't the only person that day who was taken with Pete.

Later that evening, one of the reporters who'd sat down with him pulled me aside at the lobby bar. "Lis, seriously. I feel like he's wasting his time running for DNC chair . . . he should run for president."

I laughed.

"No! I'm serious—he's like JFK. He's like Obama. He's a once-in-a-generation talent. You've found *it*." I didn't say much in response—it seemed like a lot to process. *JFK? Obama?* We ordered another round of drinks. But deep down, I felt what she was saying, and the seed was planted in my mind.

He was the total package: he combined raw communications talent with really shrewd political judgment. The weekend of Trump's inauguration, Pete—along with the other chair candidates—was invited to address a gathering of the Democratic Party's biggest donors hosted by David Brock, the very-wired founder of Media Matters for America. Some of the biggest money people in the Democratic Party would be in attendance, providing valuable exposure for a new face like Pete. The only problem? It fell the same day as the first annual Women's March, which would be taking place in cities across America. We collectively made the decision that Pete would skip the donor event and attend the South Bend march. He was the only candidate to make that calculation, which, in retrospect, is a little scary. He'd frequently raise it at forums, incensing his opponents. They knew that they looked like fools for missing the biggest organizing event in decades—all in the name of buttering up big donors.

The following weekend, we were in Houston for yet another DNC meeting. That Friday, Trump signed into law Executive Order 13769, more colloquially known as the "Muslim Ban," as it banned admission to the United States from several predominantly Muslim countries. Almost immediately, organic protests began to form at international airports across the country, including at George Bush Intercontinental in Houston.

We got wind of the protest while schmoozing at a DNC cocktail party, and once again, we faced a split screen and choice between siding with donors over room temperature white wine or standing with grassroots activists. The choice was clear for us. Still, we didn't want to tip off our opponents to our plans, so we devised a staggered exit plan whereby Pete would leave the event first, followed minutes later by the rest of us.

I was gabbing with the senior adviser to Tom Perez—future DNC chair Tom Perez—when, out of the corner of my eye, I saw Pete heading toward the exit. I waited a minute or so, told the Perez staffer that I had to go to the bathroom, and asked him to grab me a glass of wine. I had no plans to drink it. Instead, I walked straight to the door and within thirty seconds found myself piling into a rented minivan with Pete and a few others.

Just a few minutes into the drive to the airport, Pete announced, "So I'm going to tweet that we're on our way to the Houston airport protest." Pete's press secretary, Matt Corridoni, and I erupted in screams so dramatic that you'd have thought that Pete had threatened to kill a puppy. We told him that under no circumstances could he alert our opponents to our plans—at least not until we were closer to the airport.

The rush-hour traffic was brutal. After twenty minutes inching along the highway, I got a text from the Perez staffer that I'd ditched to "go to the bathroom." By that time, he'd put two and two together: "You motherfucker! Well played, well played." Apparently, soon after Pete and his staff had all disappeared from the donor event, the other candidates realized something was up and decided to hop in their own rides to the airport.

So Pete found himself in the middle of a surreal scene at the Houston airport, addressing a crowd of hundreds of protestors in the arrivals terminal. After the chair candidates—who by that time had also reached the airport—spoke, a tired-looking Middle Eastern family emerged from customs to roars from the crowd.

The ad-hoc group of ACLU attorneys that had assembled to offer legal advice to travelers affected by the ban quickly hit a

Lis Smith

language barrier. Word spread throughout the crowd that the family was Iranian: "Can anyone speak Farsi?" Pete raised his hand—he'd learned Dari, a sister dialect during his time in Afghanistan—"I can." And so we were ushered to where the family was standing.

It was clear that something was amiss as he began trying to speak with them. Finally, after a few fits and starts, he had an aha moment and he and the father began conversing comfortably. In the middle of all of this, a Univision television crew broke through and tried to interview the father in Spanish. I watched Pete translate the reporter's questions to the father then relay his answers back to the TV reporters in Spanish.

After the interview, Pete asked his assembled staff to help him get the family to the parking lot. He pushed their baggage cart, talking with the father the whole way to their car.

Once we had packed them up and sent them on their way, he told us the full story. "So that was weird," he said incredulously. "I went up, thinking they were Iranian, and started speaking Dari to them, but it was clear they didn't understand. Finally I put two and two together! They're not Iranian, they're Jordanian! They spoke Arabic!" He paused and shrugged. "Thank God, 'cause my Dari is a little rusty."

A little rusty? I was a little rusty when I failed my first driving test because I couldn't execute a three-point turn on an empty suburban street. I was admittedly a little rusty at my last Dartmouth reunion when I couldn't sink a Ping-Pong ball into a plastic cup of Keystone Light. On my best, most superhuman day, I couldn't even begin to fathom shifting between English, Dari, Arabic, and Spanish in one conversation. I'd worked with smart and talented people in politics—people whose talents far outstripped those of most mortals—but this was a whole new level.

Who was this guy?

The process of running for DNC chair is a bizarre one. The chair is chosen by a group of 440 political insiders—state party chairs, donors, and local activists. The way to their hearts is through their egos and pocketbooks: they want their asses kissed and demand firm financial commitments to their local party organizations. The media is largely irrelevant. That fact didn't dissuade me. I knew that Pete was an interesting subject and that reporters would jump at covering someone with his background and profile. I connected him with as many reporters as possible for interviews and profiles. He racked up clip after clip, which I then blasted out to our national press list. By the end of February, a *New York Times* reporter replied to one of my blast emails, "are u getting paid on clip commission??" It also befuddled our opponents: *Why are they doing so much media?* I just *knew* that the DNC chair race wouldn't be the end of Pete's national story.

The night before the DNC chair election in Atlanta, Jen called the core team into the war room that we'd fashioned for ourselves off the side of the lobby of the Marriott Peachtree Corners hotel. There was some bad news. That afternoon, twenty of Pete's supporters had defected to Perez—in part because Perez looked like the strongest candidate from the normie/Obama wing of the Democratic Party against Keith Ellison, who came from the Bernie wing. The other reason, which we learned that day, was that Valerie Jarrett, Barack Obama's former senior adviser, had been making calls to them to support Perez. (When Pete and Obama met after the DNC chair election, Obama confided in him: "You were too good for that job anyway.") Without those votes, Pete had no conceivable path to victory. If he stayed in the race, he wouldn't win. As a group, we made a decision: he should drop out before the vote. We called Howard Dean—Pete's highest-profile endorser—into the room and gave him a download on the situation: immediately, he agreed with our call.

Pete decided to sleep on it. He was gutted at the thought of just dropping out—especially after so many people had stuck their necks out for him by endorsing, donating to, and volunteering for him. But by the morning, we had his decision: he'd be getting out. We didn't share the information with anyone before Pete took the stage to address the DNC delegation in what was expected to be a rah-rah speech. In the large convention hall, I could see the press cordoned off in their designated area, impatiently waiting for "news" to be made.

When Pete finally took the stage, his surprise announcement was met with audible gasps. But he didn't just leave it there. He delivered a profound speech—managing to bring a room full of frenetic political operatives and scoop-hungry reporters to a standstill. You could hear a pin drop as he spoke:

> *Politics at its worst is ugly, but politics at its best is . . . magnificent. Because it's not just about policy. It is soul craft. And it is moral.*
>
> *The world is not divided into good people and bad people. No matter what we like to tell ourselves. Every one of us can do good or bad things every day. You can stick up for a friend with the utmost integrity and then go to work and do something that compromises your values. You can spend all day making your community a better place and then go home and hurt someone you love.*
>
> *We're not all good or all bad. No matter what we tell ourselves, it doesn't make you a good person or a bad person how you vote in a general election or a primary election: we're just people . . . human beings capable of doing good and bad things and that is why leadership matters most.*

The audience was caught off guard, but still rapt. This was a convention for political hacks, after all. The bar for success that day was to see who could kick Trump the hardest in the balls and at the loudest decibel. Anyone who loves partisan, political bromides should book their hotel room for the next party chair vote.

Pete's speech was like counterprogramming. His remarks were quiet, thoughtful, and philosophical. They were beautiful.

Tom Perez won the DNC chair election that day, but it felt like most of the buzz was around Pete.

Axios's Jonathan Swan tweeted: "One of the smartest Ds I know told me today that the most important outcome of the DNC race was making Pete Buttigieg a national figure." Safe to say, it was a clairvoyant observation.

A few days later, I got a call from Pete thanking me for my work on the campaign: "Lis, I know you went above and beyond. We didn't get everything right on the campaign, but you did." He was at peace with losing his first big national campaign: "I think we won by losing." I was shocked when he told me that he'd directed the campaign manager to give me an extra month of pay—I'd heard of winning bonuses on campaigns, but never a losing bonus.

I continued to work with Pete, ensuring that his name stayed in the media. He officially rehired me a couple months later. There was no express acknowledgment of why the mayor of South Bend was retaining a national, New York–based political strategist. But it was clear, to me at least, that it wasn't because he wanted to seek a third term as mayor.

To people on the outside, we made the oddest of odd political couples. Superficially, we couldn't have been more different. He was a quiet and reserved midwesterner with a penchant for quoting scripture. I was an aggressive, ostentatious New Yorker fluent in four-letter words. In meetings when he'd get too dry or wonky, I'd audibly sigh and mouth "boring" at him. Similarly, when I'd get a little too colorful, he'd roll his eyes theatrically and refocus the conversation. It might sound silly and a bit like a buddy-cop routine, but our different styles were central to our dynamic.

He benefited from having someone who could push him outside his comfort zone a bit and help him adopt more of the killer instinct that you need to succeed in national politics.

He made me better at my job, too. Working with New York politicians sometimes brought out the worst in me—the mean, the petty, the brash. Pete had no desire for me to be any of those things. He reminded me of why I got involved in politics in the first place.

We kept up a regular, ambitious schedule for Pete in terms of events and meetings with national political reporters. On the most basic, rational level, I realized that the idea of Pete—then just thirty-five—running for president was absurd. But at the same time, I kept coming back to my belief that if anyone could shock the political system, it could be him.

He was certainly in demand in red states, where traditional national Democrats couldn't connect. He keynoted Democratic dinners in Nebraska, Utah, and Kansas, and his remarks were always met with standing ovations. The rapturous reception he received from Kansas Democrats aside, our trip to Topeka was the first glimpse I'd gotten at the challenges he faced as an openly gay man on the national stage. Pulling up to the hotel where he was scheduled to address hundreds of Democrats, he was welcomed with signs held by congregants of the notoriously antigay Westboro Baptist Church: SAME-SEX MARRIAGE DOOMS NATIONS, GOD HATES FAG ENABLERS, and GOD H8S FAG MARRIAGE. As soon as we entered the lobby, we were greeted by armed security guards whom the party had hired at the last minute to provide protection *just in case*.

On August 1, 2018, Pete convened a small group of his closest advisers in South Bend. Finally, he said out loud what we'd all been thinking: he'd made up his mind, and he was going to run for president in 2020.

It was beyond audacious. Audacious doesn't even describe it. I remember looking around the room as he spoke the words out loud. I was the only person with any national campaign experience. Most of the people there had never worked outside of South Bend or Indiana.

A few months later Pete's campaign manager, Mike Schmuhl, and I met up in New York City for our first big brainstorm session. We brought completely different styles and life experiences to the table. He'd never worked in presidential politics before—the biggest race that he'd managed was Joe Donnelly's campaign for the US House seat in the South Bend area. He'd gotten to know Pete when they attended high school together, later serving as his mayoral chief of staff. He was a soft-spoken, unfailingly polite Hoosier, who would pause any conversation—no matter how big or small—to say "God bless you" if anyone within earshot sneezed. *Politico* described him as "sound[ing] like the host of classical music radio who just returned from a yoga retreat." That's not to say he was a pushover, by any means. Mike would never interrupt anyone when they were speaking, but he never let anyone speak over him if he had the floor.

Pete was putting his full confidence in the two of us to steer his campaign, so it was in everyone's interest for the two of us to get along. The big question on everyone's mind—Mike's and mine included—was whether we actually would. In those early days, we found ourselves bonding over our shared love for Toby Keith, cowboy boots, and eccentric professional athletes. We shared some deeper similarities as well: We had zero tolerance for the bullshit and con artists that run rampant in national politics. We had no desire to run an old-school, by-the-numbers campaign; Pete, Mike, and I were all born within a year of each other, and we wanted to build an operation that reflected our elder millennial sensibilities. Mike and I shared an unwavering belief in Pete—we *knew* that he could be the next Democratic nominee. And we both *got* Pete and what he was thinking. On the campaign, I'd joke that I was Pete's id, and that Mike was his heart and soul.

For that fateful first meeting, Mike picked the Rusty Knot—a now-shuttered, nautical-themed dive bar in the West Village for our four P.M. beers. His reasoning was sound—it was close to where I lived, it served Busch beer on tap, and it had a free juke-box. It was empty when we walked in, so we picked the best seat in the house—a small booth underneath a retro, neon sign depicting two buxom, topless women—their nipples tastefully crossed out with a black Sharpie. I mean, the perfect place to plan a presidential run, right?

Over the course of the next two hours—and four or five rounds of Busch—we began to hatch the outline of a campaign plan—the ideal timing for an announcement, the staff we needed to hire before launch, and policy issues that Pete should champion. As we cheered over our third round, Mike looked at me incredulously: "Pal, can you believe we're doing this? This is *wild!*" Somehow, I think that the Biden campaign's first brainstorm was a tad different.

From that first meeting through Pete's announcement, I flew out to South Bend every week or so to meet with Pete and Mike. We fell into a pretty regular rhythm—we'd hold planning meetings during the day, where we'd discuss everything from the budget and campaign structure to messaging and media strategy. At night, we'd find a way to distract ourselves. Sometimes it was playing board games over beers in Pete's living room with Pete, his now-husband Chasten, and his two rescue dogs. In a critical moment of foreshadowing for the future Transportation Secretary, we played a Saturday night game of Ticket to Ride—a board game in which players compete over who can construct the best rail routes across the United States. Pete and Mike shook their heads as I used up every turn to build a railway from New York City to Atlanta and ultimately Miami. My rationale was simple—I wanted to build a more efficient way to get to the towns I liked to party in. "Lis, you know that's not how this game works, right?" Pete asked me. I came in a very distant last place, but I stand by my choices.

Other nights, we'd hit the town. Right before Christmas, as an evening of planning was winding up, Pete announced to Mike and me with great pride that the pop-up bar trend had finally hit South Bend—"It's not New York, but it's *pretty* big for South Bend." Pete was proud of his record of revamping the city and making it an attractive destination for millennials to locate.

Pete, Chasten, Mike, and I showed up at the holiday-themed Santa's Workshop and were greeted by a large, bro-y bouncer wearing a Santa Claus costume. He clearly wasn't paying much attention as he began to check our IDs. Pete asked him cheerfully, "So what's the scene like tonight?" Without looking up from checking Mike's ID, Santa responded detachedly, "Oh, you know. It's lit in there."

"Lit?" Pete asked, turning to the rest of us, eyebrow cocked. The term was foreign to him. Sometimes he was like a seventy-year-old man in a thirty-six-year-old's body.

By the time Santa finally got to Pete's ID and his very recognizable Maltese last name, he nearly choked: "Oh my God. I didn't realize I was speaking to the mayor. Oh my God."

"Don't worry about it," Pete assured him with a chuckle as we walked in. The crowd inside was less oblivious to Pete's presence than the muscle outside. The bartender offered us free drinks, which

Pete declined, and the other revelers in the mostly twenty- and thirtyish crowd came up to politely introduce themselves to him. There was something so wholesome and refreshing about how he approached being mayor—he never traveled with throngs of staff or asked for special treatment. When people addressed him as "Mayor Buttigieg," he'd tell them to just call him "Pete." And if he got to Barnaby's—the local pizza pub—and saw a thirty-minute line for a table, he'd wait in it like anyone else.

During that period, we made a couple DC trips to meet with national reporters and potential donors. I packed his schedule with anyone who would give him time—we did rounds with the political units at the TV networks; coffee with the older, well-respected DC columnists; and beers with the younger political reporters who would be covering the presidential. Some of the meetings went well—*Oh, this guy could be interesting.* Others were downright rough. The *New York Times* was an especially tough nut for us to crack. When I finally secured a happy hour sit-down with two of its political reporters at the *Times*'s lobby restaurant—BLT Steak—Pete got a taste of the skepticism he'd face in the race ahead. Looking around at the table, one of the reporters turned to me and groaned audibly: "Lis, what are you doing? My God, he's not gonna run for president, is he? He should run for governor or Senate first. Go work for Beto." As humiliated as I felt, I can't imagine what it was like for Pete. He swallowed his pride and made his best pitch for the next hour. The put-down didn't dissuade him. If anything, it tingled his competitive instincts.

On New Year's Eve, Elizabeth Warren was the first major candidate to announce. She was followed by her Senate colleagues Kirsten Gillibrand and Kamala Harris. We initially planned for a January 14 launch, but life intervened. A few months earlier, Pete's dad, Joseph, was diagnosed with lung cancer. By January, he had been in and out of the hospital too many times to count. He was dying, and Pete—his only child—didn't want to be away from his side. Joseph was a character, and impressive to boot. He was an intellectual's intellectual—a Gramsci scholar and beloved professor at the University of Notre Dame, who spoke with a thick Maltese accent—even forty years after he'd immigrated to the United States.

It's hard to imagine the stress of caring for a dying parent while planning a long-shot presidential campaign. But Pete managed to handle it all with grace. He didn't talk about it a lot on the campaign trail—the topic was painful—but he always called campaign staffers if he learned there'd been a death or illness in their families. When my friend Kassidee's mom was in the intensive care unit at an El Paso hospital, I mentioned off-handedly that she was a Pete fan and would love a copy of his book. Early the next morning, Pete signed one for her, and Mike promptly FedEx'd it to the hospital. Kassidee stayed up all night reading it to her mom at her bedside. The next day, she passed away.

That's the thing about Pete that his detractors sometimes didn't understand. It was easy for them to deride Pete as "robotic" at times on the campaign trail—he wasn't an outwardly emotive guy. But it's not that Pete didn't feel the same—or even more intense—emotions as other politicians—he just didn't see them as fodder for performance art.

We navigated all of these complications and landed on a date for the announcement: the twenty-third of January, since it would coincide with Pete's trip to Washington, DC, for the annual US Conference of Mayors meeting. It guaranteed us a baseline of media coverage and would allow him to network with the hundreds of mayors who descended upon the confab from every corner of the country.

Still, the announcement wasn't exactly the splashy, highly produced type of event you normally associate with presidential announcements. At the time, I was just one of three paid staffers, so the task of organizing the press conference fell to me. Anyone who knows me well would tell you, politely, that logistics aren't "my thing." I can write speeches and communications plans, draft pithy and newsy statements for the press, and hold my own in rooms filled with heady and important people, but I'm useless the second you ask me for directions. I chose the location for Pete's announcement press conference by conferring with my personal god—Google Maps—furiously seeking an affordable hotel conference room within a three-block radius of the White House. It might not sound like the most sophisticated of strategies, but I knew that proximity to 1600 Pennsylvania was key. Reporters were loath to stray far—knowing that at any minute, the president could fire a cabinet member via tweet or call a last-minute press conference. I settled on the Hyatt Place on K Street, haggling them down from their initial asking price of $2,350 to $1,000. Our budget was that tight.

I slept for less than two hours the night before the announcement. I didn't have any support staff, and I was up late making sure that everything was ready for the next day—Pete would only have one opportunity to make a first impression, and I didn't want to mess it up. Before the press conference, I headed to Pete's hotel to brief him. From the couch in his room at the Marriott, I peppered him with likely questions as he ironed his white shirt at the foot of his queen bed. He didn't look up once from his task—producing the most pristinely pressed white shirt I'd ever seen in my life.

The press conference was held in a nondescript conference room. It was just Pete, a podium, two American flags, a lot of neon lights, and forty members of the barely interested national media. Through the grace of God, Donald Trump wasn't making news that day, so we technically had their attention. One of the first questions Pete received? "Mayor, for folks who may just be learning about you and watching, what's the best way to pronounce your last name?" Clearly, we had a lot of work to do on name recognition. How on earth could anyone ever learn to pronounce "Buttigieg?"

Money was an issue, too. The day of the announcement, Pete raised $120,000 online—not exactly pennies, but nothing compared to the first-day hauls of most presidential campaigns. For comparison's sake, Biden, Beto, and Senator Bernie Sanders raised $5.85M, $5.64M, and $5.26M, respectively, when they entered the race. We set what we thought was an ambitious goal of raising $1 million in the two-plus months remaining in the fundraising quarter. It would be enough to build a skeleton staff and keep the lights on through the spring.

Pete's father died five days after the announcement. He got to see his son announce for president, and Pete got to be by his side as he passed away peacefully. One of the last conversations the elder Buttigieg had with his son was to tell him how proud he was of him.

ELEVEN

The Long Shot

So—no name recognition. No money. No shot, right? Not exactly.

That's where I came into the equation. I had a theory of the media and an effective media strategy that could help an unknown presidential candidate break through. The nature of the news media and media consumption was changing—and much faster than any of our opponents realized. The traditional gatekeepers were dead—I'd learned that in 2012 through a lot of trial and error. But still, Pete's opponents were campaigning like it was 2004. They limited access to the media, thinking that it would add to their mystique and "presidential" aura. *Good luck with that strategy in the largest presidential field in modern times.*

To the extent the other candidates talked to the media, it was softball appearances on MSNBC or interviews with the *New York Times* or maybe, occasionally, a cameo on a Sunday show. I have no doubt that anyone advising them to take this tack did it in good faith, but they were working off of a tired playbook. Don't get me wrong—a good appearance on *Morning Joe* or *Maddow* could juice contributions and stoke interest in a candidate—*if* they had something interesting to say. I'll get to that in a second. But this was 2019—we were living under a political culture defined by Donald Trump's "no fucks given" media approach in 2015 and 2016. The guy would basically talk to anyone—he leveraged his celebrity to get on sports and entertainment outlets, dominating them even more than he did the political press. It was a better fit for him than the traditional media, but still, he was a regular call-in guest on CNN and MSNBC. He sat down with the *Times* and the *Washington Post* with more gusto and availability than his opponents. He was as fearless as he was unprepared; he completely bombed some of the interviews, like his sit-down with the *Washington Post* editorial board, where he equated the economic threat posed by Mexico with China's and devoted more than thirty sentences to defending the size of his hands: "My hands are fine. You know, my hands are normal. Slightly large, actually. In fact, I buy a slightly smaller than large glove, okay? . . . I don't want people to go around thinking that I have a problem."

In any normal world, comments like that to an editorial board would dominate coverage for days. So small—no pun intended—but so unserious. But no one remembers that interview today—no one remembered it a day after it was given. Why? Because Trump never ever let the news cycle settle on one narrative about him or calcify around one comment he gave. As soon as the horrifying transcript of his conversation with the *Post* was published, he was flooding the zone with other equally terrifying comments. *You think that was grotesque? Let me show you grotesque.*

Even after Trump won, I think that a lot of Democratic strategists watched the Trump media strategy and recoiled—*we can't do this! This is so tawdry!* To some extent, I'd agree with them. No one should emulate Trump, *but* they should try to learn from him. Twenty-point policy plans leaked to the *New York Times* don't make the evening news, nor do carefully crafted talking points. If you're only talking to an audience who will look up online the details of your medical device tax policy, you've lost the plot. You've gotta meet people where they are—through the outlets they trust. And you've gotta do it on their own terms.

I'd studied the Trump approach to the media, as well as Hillary Clinton's during the same election. As Trump caused media tsunamis on top of media hurricanes, earthquakes, and tornadoes—Clinton went over 250 days between December 2015 and August 2016 without a press conference. When she did interact with the press, they'd latch onto any "gaffes" she made—giving them wide coverage and letting them penetrate with voters until the next time she spoke publicly. She didn't understand the cardinal rule of media: if you don't feed the beast, it will feed on you.

While I had grudging respect for Trump's feral brilliance, I looked to McCain's 2000 campaign as a better model. Through sheer force of will and the savvy model of the Straight Talk Express, his all-access, on-the-record, no-holds-barred media bus, McCain made himself into a fearsome competitor for the Republican nomination against George W. Bush. Bush was the son of the last Republican president. He raised $68 million in 1999 to McCain's $15 million. He dominated two of the three lanes that you needed to win the nomination and the presidency—money and name recognition. McCain, however, owned the third—the media.

Early on, Pete and I made the decision to follow the McCain route. McCain was brash, chummy, and colorful with a notoriously short fuse. Basically the opposite of Pete. On calls, Pete—whether angry or pleased—would respond to ideas thrown at him with a simple: "Hm." Over time, I learned to decipher his language: "Hm" was bad or unimpressed. "Hmmmm" was good. It was the Pete equivalent of saying: "I'm intrigued. Tell me more."

I put him in front of every reporter who had time for him. I offered news outlets behind-the-scenes, all access to him. VICE News embedded themselves on Pete's first trip to Iowa. CNN ran a package in February that featured at-home interviews with Pete and Chasten. In the middle of the taping, Pete's dog Buddy—an overweight one-eyed rescue puggle—jumped into the shot and interrupted the interview by barking at Kate Bolduan and the CNN cameraman. We connected Pete with outlets that other politicians wouldn't dare to speak with: Barstool Sports, TMZ, and *Playboy*—a magazine you usually associate with naked women, not gay, scripture-quoting, midwestern mayors. We put him on the air with right-leaning media like Fox News and Hugh Hewitt, as well as far-left hosts on the Intercept. Basically, we made it pretty hard for you not to hear of the guy.

We also flooded the zone in the early states—"no" was not a word in my vocabulary when it came to any media request that came in from Iowa. There was no outlet too small that he could speak to in my estimation, and he would sometimes roll his eyes at me as he found himself on a nine P.M. call with a nineteen-year-old Iowa college student who couldn't figure out how to turn on the recorder.

As I mentioned before, Pete had the raw skills to handle the press strategy. His tone and style were completely refreshing. I'd seen his ability to consume information and news quickly during the DNC chair race—but I knew the presidential race was a whole different ballgame. Still, even when I threw out tough questions on issues we'd never discussed, it was like he always had a perfectly formulated, thoughtful, and oftentimes unconventional answer ready. He was like a major league hitter following the rotation of a baseball hurtling at 100 mph toward him.

Before the exploratory committee announcement in DC, Pete, Mike, and I had combed through every single detail of the day, including what Pete should wear. Pete wanted to go with one of his ill-fitting suits, but I was adamant that he should lean into his youthful, mayoral persona, ditch the suit jacket, and develop a distinct look—rolled-up white shirtsleeves and a blue tie. He was in a historically large primary and needed to establish a distinct visual brand—one that people could immediately identify with him.

It's telling about the nature of modern society and politics in general that it was one of the more controversial decisions we made. I heard from his Harvard friends and from big-time political consultants about what a mistake it was: "You realize he looks younger, not older, right?" Pete, to his credit, stuck with it.

It's a small thing seemingly, right? What a politician wears day to day absolutely matters less than their leadership qualities or positions on issues. But it's the height of naivete to think these things don't matter. In 2008, Barack Obama's decision to forgo a tie with suits signaled his effortless cool and outside-the-beltway sensibilities. Hillary Clinton's pantsuits became a feminist rallying cry. Look at Bernie Sanders, whose white hair and rumpled wardrobe that looked like it had been—minutes before—scavenged out of a bin at a church giveaway became iconic. He wasn't trying to be a *GQ* god, and it worked in his favor.

Pete's look quickly became a thing. When we were in New York City, people would stop him on the street: "I recognized the white shirt and blue tie!" Pete was one of the first 2020 Democratic candidates to get the *Saturday Night Live* treatment. When the actor Paul Rudd was guest hosting, he played Pete in a skit that I watched from an Applebee's in Manchester, New Hampshire. Rudd parodied a Pete appearance on *The View*:

I'm ready to work!

See my exposed forearms? I may only be 37 years old, but I do feel like I represent everyday Americans. I'm just a Harvard-educated, multilingual, war veteran Rhodes Scholar. I'm just like you!

Pete was a good sport about it all. Before events and media appearances, he'd ask me, "So am I wearing the 'uniform' today?" He'd internalized the power of creating an image. In our merch store, we sold a kid's T-shirt—white emblazoned with a blue tie—and we couldn't keep it in stock. It became a Halloween costume that was so identifiable that Pete's doppelgänger—Brad Stevens, the wunderkind head coach of the Boston Celtics—could dress up in it and set off a Twitter meltdown among the sports and political media.

Always, always, it felt like the "adults" were looking down at us from their perch of supposed wisdom. Months before Pete announced his presidential exploratory committee, we'd briefly bantered about the Supreme Court. It's an issue where I don't have particularly nuanced views: I think the modern-day court is an embarrassment. The idea that these political appointees—some exceptionally brilliant, some less smart, some younger, some in their eighties on their death bed—could be seen as the ultimate arbiters of the law is obscene to me.

At an event a few weeks after Pete announced his exploratory committee, he was asked about the concept of "court packing." He responded immediately that he was open to the idea: "It's no more a departure from norms than what the Republicans did to get the judiciary to the place it is today."

Watching the event live, I knew immediately that it was "news." When I tweeted out Pete's line, it went viral. Prominent progressive types didn't think that Pete had much of a future, but they were positively giddy that a presidential candidate had opened the door to reforming the Supreme Court. *This* was fresh thinking, the sort of truth telling that could come only from a thirty-seven-year-old candidate.

Pete's mentors and outside advisers clearly disagreed. The next day, the emails, texts, and calls started to roll in.

SUBJECT LINE: Court Packing
TEXT: Does he support this???

My personal favorite was the guy who called to tell me: "This is catastrophically stupid of him! Has he ever read about FDR?!?!?"

There were a lot of charges that one could level against a guy like Pete—but not knowing about motherfucking FDR and his legacy wasn't one of them. More importantly, what does a political battle that transpired under FDR in the 1940s have to do with a conversation in 2019?

It was an early and small test—one that transpired outside of the glare of the media because the media didn't care about Pete then. He listened to the detractors but ultimately rolled his eyes at them. He wasn't going to be the safe, boomer candidate they wanted. He dug in his heels and was the first presidential candidate to unveil a plan to revamp the Supreme Court. Eighteen months later when Ruth Bader Ginsburg passed away and court expansion became Democratic Party gospel, it was hard not to gloat to the early naysayers.

Our media strategy was paying off.

A week after his exploratory committee announcement, Pete wowed the panel and crowd during an appearance on *The View*. The hosts, who usually didn't agree on much, ended the interview with effusive praise for Pete. We hopped in the car to head over to MSNBC for an off-the-record editorial meeting with Phil Griffin, the head of the network, and a couple dozen producers and bookers. Five minutes into Pete's presentation, Phil's eyes widened as he brought his hand to his mouth to hide the smile forming on his face. He was having the same reaction I'd had the first time I sat down with Pete: *Holy SHIT! This guy is good!*

Pete was winning friends, but he needed a breakthrough moment. He'd soon get it.

The moment that changed everything—a CNN town hall—finally came on March 10, forty-six days after Pete launched his exploratory committee.

The CNN town halls were an important and welcome innovation in the 2020 campaign. The concept of a town hall wasn't radical—it's a staple of campaigning at every level. What CNN did differently was cede a full hour of primetime coverage to candidates, allowing them to showcase their talents and ideas—or lack thereof—to the American public. If you were a candidate who thrived on canned answers and could just barely squeak by in a five- to seven-minute cable news hit, the town hall format was a nightmare. If, on the other hand, you were a candidate like Pete—someone who knew who he was, what he believed in, and how to communicate it in a compelling way—the format was a dream.

For weeks, I harassed CNN nonstop for a town hall for Pete. In January and February, they'd hosted town halls with Senators Sanders, Harris, and Klobuchar. Pete was, understandably, a bit farther down their list.

Finally, CNN approached us with an offer. They were going to host a town hall marathon on March 10 from the South by

Southwest festival in Austin, Texas, and they were offering slots that evening to former Maryland congressman John Delaney, Congresswoman Tulsi Gabbard, and former San Antonio mayor and HUD secretary Julián Castro. It had strong B-side, undercard vibes, but we knew we didn't exactly have a ton of negotiating power and agreed to it.

I was caught by surprise when CNN publicly announced the town halls a week or so later. Castro's name was nowhere to be found. I learned that he'd balked at the prospect of sharing a night with three other presidential candidates and pulled out—hoping for his own solo night. Even then I knew that was a mistake—it was early in the campaign, anything could happen—especially with a free hour of national TV.

We knew we had to make the most of this moment. So we gathered our small campaign team—it had grown to seven whole staffers at that point—and set up an operation that would allow us to clip noteworthy moments as they happened and share them with our most die-hard supporters. We set up lists for "rapid response"— essentially a list of supporters who could share clips on their social media accounts, and also for our thousands of grassroots donors who could help us use the moments to raise money online.

In the days leading up to the town hall, Pete started to work-shop a critique of Vice President Mike Pence in media interviews at South by Southwest. The first time I heard him deliver it, I had an aha realization. His riff still needed work, but this was a moment that could really pop for Pete. Pence was one of the left's favorite villains—he virulently opposed LGBTQ rights and a woman's right to choose. He was also a natural target for Pete—he'd served as governor of Indiana during Pete's early tenure as mayor. The day of the town hall, we held a final prep session where Pete honed his answers, including a potential Pence smackdown.

Pete and four of us staffers watched Delaney's and Gabbard's town halls backstage in a cramped dressing room in Austin. The only notable moment that came out of either of them was Gabbard's refusal to denounce Syrian president Bashar al-Assad's gassing of his own people. Suffice it to say, it was not a good development for her.

When Pete took the stage, he had a case of nerves. He spoke a little too quickly and ran through answers without really playing to the crowd in the hall. After a few questions, he began to settle in. He worked the stage and let his answers breathe.

Halfway into the town hall, a question came from a town hall attendee. It was about Mike Pence:

"Hi. Thank you for your service and your position in politics. The question I'd like to ask you is, as you pointed out, Vice President Pence is obviously quite conservative. And in regard to these conservative views, in regard to religion and in sexuality, in comparison to the average voter or the voter in Indiana, let's say, are his views an aberration? Or is this really representative of the state? Or are most people more like you in your more liberal views about us as humans?"

Mike and I looked at each other with anticipation—here was the moment we had been waiting for.

Pete started out his answer, "Please don't judge my state on Mike Pence . . ." The CNN crowd laughed and Pete gave an impassioned answer about all the disagreements he had with Pence. But he didn't quite stick the landing. As his answer started to wind down, the CNN moderator Jake Tapper jumped in with a follow-up question, "Do you think Vice President Pence would be a better or worse president than Donald Trump?"

Pete physically recoiled at the question and took a beat as horror theatrically flooded his face at the Sophie's Choice that Tapper presented him. The audience exploded with laughter.

"Does it have to be between those two?" Pete asked Tapper with a devilish smile on his face. Tapper responded, "Politics is about choices, man! You know that." Pete responded:

> *I mean, I don't know. It's really strange. Because I used to at least believe that [Pence] believed in our—I've disagreed with him ferociously on these things, but I thought, well, at least he believes in our institutions and he's not personally corrupt.*
>
> *But then—but then how could he get on board with this presidency? How could somebody who—you know, his interpretation of scripture is pretty different from mine to begin with. OK, my understanding of scripture is that it is about protecting the stranger and the prisoner and the poor person and that idea of welcome.*
>
> *That's what I get in the gospel when I'm in church. And his has a lot more to do with sexuality and, I don't know, a certain view of rectitude. But even if you buy into that, how could he allow himself to become the cheerleader of the porn star presidency? Is that he—is that he stopped believing in scripture when he started believing in Donald Trump? I don't know. I don't know.*

In slow motion, Mike and I could feel something happening. Tapper's face and body language changed as Pete navigated his way through the response—he could spot a good TV moment when it was happening. It was just a master class in how to give an answer to a question—Pete effortlessly combined his gentle, aw-shucks demeanor, with a GIF-able facial expression, with a reference to scripture, and—finally—with a nut cutter of a line about Pence and his adherence to his own faith. (The line clearly got under Pence's skin. A few weeks later, he fired back at Pete in an interview—a dream scenario for an underdog candidate.)

In politics, you always hope for a *moment*. Most of the time it never comes. Even winning campaigns often struggle to pinpoint the exact second things changed for them—not in this case. There was no ambiguity here. Our website was flooded with traffic and donations. Twitter was alight with mentions of Pete. Our phones were inundated with messages from previous Pete skeptics singing his praises. When he picked up his iPhone after the town hall, he said under his breath, "Wow, Oprah just texted me."

We woke up the next morning to a whole new world. The news organizations that had ignored Pete were suddenly all about him. The *Washington Post* wanted to embed a reporter with him. Fox News and MSNBC reached out about hosting their own town halls with him. Mass-market shows like *The Ellen Show*, which had previously rejected our outreach, were suddenly all about him.

To date, our biggest fundraising day had been the day of Pete's launch. We were stressed that we wouldn't be able to meet our $1 million fundraising goal for the quarter. We were even more behind on meeting the sixty-five thousand donor threshold that the DNC had set as a qualification to make the first debate. The night before the CNN town hall, we had just twenty-five thousand donors.

Within twenty-four hours Pete raised $600,000 dollars. Within forty-eight hours, he raised $1 million and hit the sixty-five-thousand-donor requirement. Within a week, he had nearly a hundred thousand donors. By the end of the fundraising quarter three weeks later, he'd raised $7 million.

Overnight, Pete had become a political sensation. A poll of Iowa caucus-goers released the morning of the town hall had Pete garnering 2 percent—good enough for tenth place behind Biden, Sanders, Warren, Harris, Beto, Booker, Klobuchar, Castro, and Colorado senator Michael Bennet. An Iowa poll released just two weeks after the town hall hammered home just how *big* the moment was—it had Pete surging into third place behind Biden and Sanders, with 11 percent of the vote. *Vanity Fair* captured how most people felt: "Who the hell is Pete Buttigieg, and how is he polling in third place behind Bernie?"

It was remarkable. Mike described it as "warping" political time—it's like Pete had skipped ten of the usual steps of running for president over the course of one hour on national TV. The media had a new phenom. Move over Beto, the new "it" candidate was the wonder boy from South Bend.

Six days after the town hall, Pete appeared on *Morning Joe*. The planned ten-minute hit stretched out for over the thirty minutes as the hosts gushed over him. Mid-interview, Mika told him: "You're like the Mr. Rogers of the Democratic field." Just a few hours after it aired, Joe tweeted: "Mika and I have been overwhelmed by the reaction @PeteButtigieg got after being on the show. The only other time in twelve years that we heard from as many people about a guest was after @BarackObama appeared on Morning Joe." The sentiment was echoed by prominent columnists, late-night show hosts, and political talking heads. It was the start of Pete Mania.

After the Eliot tabloid flareup, I had felt people discounting me. *She flew close to the sun; she's done!* When I went to work on Pete's presidential campaign, I could hear the same voices singing a new tune: *Who? Has she lost her mind?* Now, here we were, cracking the code of presidential politics with the most unconventional of strategies.

For three years, I'd been completely, blissfully out of the tabloid and media spotlight. I'd needed time to get used to being a normal human being again—someone who didn't flinch at a camera flash or speak in unnaturally hushed tones because I was afraid that a dinner or shopping conversation would find its way onto a gossip site. But I had an unusual background for a presidential operative.

It was inevitable that I'd start to get more scrutiny. The weekend Pete announced for president, *New York* magazine ran a cover story profile written by Olivia Nuzzi. There was a section in the story about me: "Hiring Lis Smith, now his campaign spokeswoman, for the DNC race was a kind of aggro move for Buttigieg. She is well known and well liked by the national media but disliked by many of her fellow Democratic stagehands, in that particular way a certain kind of woman often finds herself disliked."

After the story posted online, friends and colleagues reached out to me offended on my behalf: "What does that even mean?" I knew exactly what it meant. Olivia, like me, had experienced her time in the barrel of the tabloids. She understood the pettiness that consumed the DC political and media class.

A couple days later, when Pete returned to the set of *Morning Joe*, Mika beckoned me over for a chat during a commercial break.

"Lis, can I give you some advice? Can you handle some tough feedback?" I assured her I could.

She launched into it: "So you had this great tweet over the weekend. Joe retweeted it, and I was going to retweet it too. But when I looked at your profile photo, it looked like you were wearing lingerie." She wasn't wrong—my photo at the time was from a good friend's wedding, where I was wearing a dress from Fleur du Mal, a boudoir-inspired brand. Safe to say, it didn't convey thoughtful, super-serious political adviser.

"He could go all the way to the White House," Mika said, pointing at Pete, then she turned to me and said, "and you're going to be the one to take him there, but not with this." As she said "this," she drew a square in the air around me. "*This* doesn't fit with Pete."

Again, she wasn't wrong. Pete was on set in his austere uniform of a white shirt with rolled-up sleeves and his signature blue tie. I was wearing a black Stella McCartney suit, a skull T-shirt, and leopard-print high heels. We weren't exactly in sync image-wise.

Mika continued, "You need to look the part. Ditch the high heels; drop the glamour act. Trust me, there will be plenty of opportunities for you to play the sex kitten after the campaign, but now's not the time."

I nodded along with the advice, but I'd be lying if I said the words didn't sting a bit. Like a lot of women working their way up the professional ladder, I associated a glamorous, aspirational wardrobe with success. I subscribed to the idea that designer heels and sleek dresses gave me authority. They probably did in a place like New York City, but Pete wasn't running for office in New York City.

"You need to go out, get some hiking boots, tight pants that show off your butt, and cute tops," Mika told me. "And you need to change your Twitter pic."

Within fifteen minutes of leaving the set, I changed my profile photo to one of me somberly advising Pete backstage at an event. I went to a nearby department store and asked the shoe salesman to help me find the best low-heeled boots they had in stock. I bought a few more casual staples and developed a campaign wardrobe.

Don't get me wrong—I still kept my edge. I'd wear leather pants, lots of vintage tees, and the occasional high heels, but I spent less time trying to look like an extra from *The Devil Wears Prada*.

There are some people who would hear Mika's words and think they were sexist—*women should wear what they want!* But her advice got to a deeper point that is sometimes hard to absorb as a woman in the workplace. We can delude ourselves and tell ourselves a fairy tale—one where we can dress and look however we please without judgment and somehow still reach the top. After all, the men who are usually celebrated in politics are slobby, disheveled, and unattractive. As Marlo Stanfield, the psychopathic drug-dealer kingpin in *The Wire*, put it best: "You want it to be one way. But it's the other way." The other way, as I learned, was feeling more comfortable in my own skin.

Little changes sometimes make the biggest differences: my feet didn't ache all the time. I spent less time at the dry cleaner. In meetings, men stopped commenting on my dresses or my shoes—they just listened to me. Not everyone was a fan. I think my mom died a little inside every time she saw a photo of me on Twitter where I was wearing cowboy boots and jeans. She thought her little girl should be wearing glamorous dresses and stiletto heels.

A few months later after that interaction, I returned to *Morning Joe* with Pete. As he sat down on set, Mika looked across the room for me. When I waved at her, she looked me up and down from afar, squinting. I was wearing tight jeans, Isabel Marant wedge sneakers, and an oversized vintage Cincinnati Bengals sweatshirt. She smiled, nodded approvingly, and gave me two thumbs-up. *Mission accomplished.*

By the time Pete formally announced his campaign in mid-April, it was undeniable: he was a top-tier presidential candidate. On a freezing cold, rainy day, we crammed four thousand people in the old Studebaker Factory in South Bend, while another fifteen hundred people stood outside in the truly miserable weather. We raised $2 million that day, and another $1 million the following day.

When Pete returned to Iowa for the first time as an official candidate, more than fifteen hundred people showed up to his first event in Des Moines. It was the largest Iowa crowd for any candidate yet that cycle, and it dwarfed the twenty people who had shown up to a coffee shop to see Pete just two months earlier.

It wasn't all pretty though. The event was interrupted by Randall Terry, a notorious right-wing activist who'd made his name by picketing abortion clinics. He and a few of his followers disrupted Pete's remarks, chanting, "Remember Sodom and Gomorrah!" The crowd drowned them out with chants of "PETE! PETE! PETE!" After they were escorted from the site. Pete took a deep breath.

"The good news is, the condition of my soul is in the hands of God, but the Iowa caucuses are up to you."

The crowd went wild. He had cleverly and gracefully handled an ugly moment. It was tailor-made for TV and a perfect split-screen moment with Donald Trump's treatment of protestors.

The next day, the protestors were back, this time with a little more flair. When I arrived at the home where Pete would be holding a backyard town hall, I surveyed the scene. I quickly saw that not only had the protestors upgraded to using a sound system, but they also were wearing elaborate costumes—one dressed as Pete, another dressed as Jesus on the cross, and yet another dressed as the devil. "Pete" was whipping "Jesus," as "Satan" urged him on over the loudspeaker. *Seriously, what the Hell?* File it away under "things you will see only at a backyard event in Iowa during a presidential election."

There was no way we could let them speak over Pete, but there was one problem: we had anticipated an intimate gathering and hadn't procured a sound system of our own. I made the call with a few of the other staffers on the ground—we'd delay the start of the event by thirty to forty-five minutes, the time it would take to procure adequate audio equipment in the small town.

By this time, Pete had arrived and was greeting political VIPs in the living room of the house—oblivious to the scene unfolding out back. I went inside to let him know the event would be running behind. I pulled him into the den, where we could speak in confidence, and filled him in on the details.

"So . . . the protestors from yesterday are back," I told him as his brow furrowed. "This is the good news: you managed to bring Jesus Christ and Satan together. The bad news is that they're both heckling you."

He busted out laughing. For the first time that morning, I started to laugh as well, then assured him we'd have a sound system soon.

Pete went on to address the crowd, plus the local and national media, with aplomb. The Iowa trip was an unambiguous success.

It felt like a fairy tale—everything was just going so well. Pete empowered me to take risks with the media on his behalf, and they kept paying off. One of the earlier and more controversial decisions we made was to engage with Fox News. It may seem hard to believe now—especially after Pete earned the moniker of "Slayer Pete" for his masterful takedowns of bad-faith arguments from Fox anchors during the general election, but we took a lot of heat for the decision during the primary.

Fox News executives had been some of the first to reach out after the CNN town hall, offering Pete one of his own on their network. Before we committed to it, I began to get inundated with calls and emails from Chris Wallace's bookers for his show the upcoming Sunday. It was an intriguing, yet also terrifying, prospect. Wallace is known to be the toughest interviewer in political news. He had yet to interview a 2020 Democratic contender, probably for that reason.

The cautious political operative in me was mildly terrified. The cocky competitor, however, was intrigued. I was frank with Pete: "This could be a big moment for you, but it could also be a disaster. Still, I think you should do it." He didn't push back—he was also competitive, and knew that for his campaign to go anywhere that he had to take some risks: "Nothing ventured, nothing gained, I'm in."

And so Pete and I found ourselves in the greenroom at Fox News's studios in Washington, DC, early on Sunday morning. He got his makeup done next to then–White House budget director Mick Mulvaney. As we watched Wallace eviscerate Mulvaney, we wondered what we had gotten ourselves into. Pete lightened the moment by telling me how one of his friends who knew Mulvaney had told him that Mulvaney embodied the famous quote attributed to everyone from George Orwell to Coco Chanel: that "you're born with the face you have, but you get the face you deserve at 50." *Ouch.*

Chris pulled no punches with Pete that morning. It was the toughest interview he'd faced to date, and the first time that he'd faced the "front-runner" treatment, even though he was far from a front-runner. Chris grilled him on crime and employment stats under his tenure as mayor, and it was the first time I'd seen Pete put on the defensive. All in all, though, it was a good interview for Pete—he'd acquitted himself well, signaling to political insiders that his candidacy was more than a flash in the pan.

Fox announced their town hall with Pete in late April—and his opponents seemed largely okay with it until Pete's support kept building. The week before Pete's scheduled town hall with Fox, Elizabeth Warren announced that she wouldn't do one with them, calling Fox a "hate-for-profit racket." It would be a beautiful, principled stance if she hadn't herself spent a lot of time on Fox News as a US senator. Kamala Harris's campaign followed suit. There were calls from groups like Media Matters for America for Pete to back out. It raised the stakes for Pete—all eyes would be on his town hall and how he handled it.

None of it mattered once Pete took the stage. Once again, he delivered a masterful political performance. Before the first commercial break, he addressed the elephant in the room. He called out Tucker Carlson and Laura Ingraham by name, quoting their latest, heinous comments.

When the show cut to commercial, I walked out to the set to

gauge the mood of the crowd. It was buzzing with excitement as a Fox exec walked over to me, laughing: "I'm already hearing about that one," he said. I shrugged faux apologetically and smiled. Just as we would never expect Fox to give us questions in advance, they couldn't expect us to preview Pete's comments. Plus, it was a win-win for everyone—Pete's comments lit up social media, and he began to trend nationally. The rest of the town hall went swimmingly.

As Chris announced that the town hall was wrapping up, the surprise on his face was palpable when the two hundred or so people in attendance jumped to their feet—"Wow! A standing ovation!" he declared.

It was a home-run, made-for-TV moment. When I walked out to the set to check in on Pete and thank Chris for moderating, Chris confided in me, "You've got a real contender here. That felt like being in a boxing ring with Mike Tyson."

The next day, the clips from the town hall were everywhere. They were featured not just on Fox, but in every hour of coverage on rival stations like CNN and MSNBC, who gleefully replayed Pete's denunciation of Laura Ingraham and Tucker Carlson for their viewers. Pete's performance received praise across the media—there were multiple stories and columns in both the *New York Times* and *Washington Post*, along with other major newspapers. It was picked up by celebrity news sites like *People* and TMZ. And it got ample play on late-night shows like the *Daily Show* and the *Colbert Show*. Everywhere you turned, there was Pete.

The town hall itself attracted 1.1 million viewers—four to five times the audience that we would get with a town hall on CNN or MSNBC. More importantly, we got millions upon millions more eyeballs from the clips that were picked up elsewhere. It reinforced our general view of the media and media mentions generally—it was worth taking risks if the strategy forced Pete into the national conversation. Even if it was Fox hosts trashing Pete, they were helping to improve his name ID and contribute to his buzz. We dominated the news cycle for days because of it and earned coverage that other candidates would have killed for.

It was a teachable moment, and one that I hope more Democratic campaigns take to heart in the years ahead.

First, re: the morality of appearing on Fox News—*Get. Over. It.* Saying you're going to ignore the largest cable audience in America because you're so politically pure is the political equivalent of burying your head in the sand. Fox wasn't going anywhere. Do you think that when Kamala Harris and Elizabeth Warren decided not to go on that somehow Fox was going to go bankrupt?

Second, there's a way to do Fox News and not get caught up in any of their bullshit. In one hour, Pete dissed two of their biggest hosts, took on the Republican president, and knocked down right-wing talking points about abortion. It's possible to go on right-leaning shows and outlets and still present a strong progressive case.

Third, how on earth do people think Fox viewers will ever hear the Democratic message or identify with it if we don't appear there? The point we made within Pete's campaign was that some of the hosts and powers that be there might be acting in bad faith, but the viewers who tune in aren't. And it's fundamentally unfair both to them and to us if we turn up our noses and refuse to talk to them. It's impossible to pierce the right-wing bubble if you just ignore it.

We continued to engage with Fox throughout the cycle. Eight days before the Iowa caucuses, Pete was the only presidential candidate to hold a second Fox town hall. Once again, the reception was thunderous and overwhelmingly positive. Our organizers in red, rural areas of Iowa recounted to me how they were able to fill up slots for precinct captains with Republicans and independents who walked into their offices after watching the town hall. The night of the caucuses, they heard over and over again from caucus-goers that Pete's town hall was what made them caucus for a Democrat for the first time in their lives.

There is not a doubt in my mind that that final appearance made the difference for Pete in Iowa. Nothing ventured, nothing gained, indeed.

Three weeks to the day after the May Fox town hall, I woke up to an early-Sunday-morning text from Mike. There had been a fatal police shooting overnight in South Bend—the police officer was white, the man that he'd killed—Eric Logan—was Black.

The news was tragic. The political implications were clear: really, really bad. Just over four years earlier, I'd seen how this scenario had played out on O'Malley's campaign. This was going to be ugly.

Immediately, we pulled Pete off the campaign trail to let him handle the very important business at home. In the coming days, he was scheduled to do a splashy West Coast fundraising swing, with big-money events cohosted by celebrities like Katy Perry. We canceled them all.

As soon as news of the shooting broke, it became fodder in the 2020 campaign. Over the coming days, national news organizations sent reporters to South Bend to cover the fallout. The headlines were rough, with the AP cutting to the chase: "Police Shooting Poses Buttigieg's Biggest 2020 Challenge Yet." And of course our opponents weren't going to let this moment pass—three days after the shooting, a reporter sent us an opposition research file that one of the other campaigns was shopping around to reporters: "Officers Involved in South Bend Police Shooting

Have History of Misconduct Allegations; Buttigieg's Police Force Plagued with Problems."

Each day brought a new challenge and new criticism. Pete took it all head-on and insisted that he would hold a public town hall meeting a week after the shooting to address South Benders' concerns. It was beyond risky—there was no way he could know how it would go, but he was planning to face the music in plain view of the local and national media.

The town hall fell on a Sunday in June when my family was holding a belated Father's Day celebration for my dad. Up until that point, I'd largely been at peace with the Faustian bargain that I'd made to succeed in my chosen career. I could decline invitations to weddings or birthdays of close friends, but my red line was blowing off anything that had to do with my dad. Between January 2017, when I'd watched Pete announce for DNC chair from my dad's hospital room, and June 2019, my dad had suffered a stroke that paralyzed one side of his body. He'd moved into an assisted living facility where he was confined to a wheelchair and received round-the-clock care. On top of the Parkinson's and the stroke, he'd begun to suffer from dementia.

The day was extremely chaotic. I was filled with guilt about not being in South Bend and was running late to the Father's Day celebration because I was helping the staff on the ground in Indiana prepare for the town hall. By the time I began the drive out to the suburbs, the town hall was starting. I streamed the audio as I sped along the West Side Highway.

It got ugly almost immediately. I couldn't see the video, but I could hear Pete being repeatedly cut off by protestors. As I turned off the exit for Bronxville and hit a red light, I texted Chris Meagher, Pete's national press secretary, "Is it as bad as it sounds?" He responded instantaneously, "It's worse." He wasn't kidding. I felt sick by the time I got to my brother's house and turned on the TV—MSNBC, CNN, and Fox News were all covering the event live.

The reviews of Pete's performance quickly flooded in. David Axelrod wrote a piece for CNN about how Pete had missed his moment to show Black Americans how he felt their pain. Axios declared it a "nightmare scenario." Still, the feedback wasn't all bad—on *Morning Joe* the next morning, Al Sharpton acknowledged that the decision was "politically risky, but [it] was the right thing to address the community."

In the coming days, I would second- and triple-guess the wisdom of Pete holding the town hall, and my decision to uncritically go along with Pete's edict and invite national reporters. He got through it, sure, but I questioned it for the rest of the campaign and wondered whether I should've pushed back harder. After all, I was on the campaign to manage the media and handle the visuals.

It haunted me even after the campaign ended. Had our handling of this fraught moment doomed Pete with Black voters? But I reconsidered that thinking in May 2020, when I saw how the city of Minneapolis had devolved into riots after local officials were slow to engage publicly after George Floyd's death was caught on tape. It might not have been pretty, but Pete had given Black residents of South Bend a hearing. It was a cathartic moment for the community. And while it certainly didn't fix race relations in his city, his decision to hold a town hall confronting the issue head-on was likely why we never saw images of buildings and cars burning in South Bend like those we saw from Minneapolis.

The first presidential debate was going to be held in Miami on June 27. We'd gotten word from a reliable source that Kamala Harris was likely to go after Pete in the debate for his handling of policing issues in South Bend. It seemed like an odd line of attack, given her own vulnerabilities from her time as attorney general and district attorney, but we couldn't take any chances. At the very least, we *knew* the moderators would ask Pete about it.

The day before the debate, Pete was falling flat in prep. It was clear that the shooting was weighing heavily on him. He'd lost the joy and confidence that had gotten him this far. I called for a break in the session, and the team conferred and agreed that without a serious change in course, Pete could have a disastrous debate. The best-case scenario was that he would just completely fade into the background. I tasked the two senior men in the room to deliver that message to Pete when he returned.

The problem? They didn't. They offered polite critiques, not the blunt assessment we had agreed upon just minutes earlier.

I had a group text chain with the two other women in the room. Watching this unfold, one of them texted: "WHAT THE FUCK!!!" I would have to take matters into my own hands. I pulled up my chair just a few feet from Pete and looked him in the eyes.

"So, this is the deal—this is a make-or-break moment for you. I'm not gonna sugarcoat that. And you cannot go up there onstage tomorrow and be Peter fucking Pan. Because Democrats are looking for someone who can beat Donald Trump, and no one thinks Peter Pan can do that."

The room went silent as Pete absorbed what I'd said. His eyes narrowed and his cheeks grew pink. "Okay, got it," he replied. "Let's get back at it." He jumped back onto the mock debate stage, and the fire that he'd been lacking suddenly emerged. Thin-skinned he was not.

The next night, things unfolded a little differently than we expected. Sure enough, the moderators asked Pete about the shooting and why the diversity of the police force hadn't improved under his watch. Pete opened his response candidly: "Because I couldn't get it done." As soon as he finished his answer, opponents like Colorado governor John Hickenlooper and Congressman Eric Swalwell jumped in to attack him. *Hickenlooper? Swalwell? Who cares? Where's the kill shot?*

Finally, Harris jumped into the fray. "As the only Black person on this stage, I would like to speak on the issue of race. . . ."

In the greenroom backstage, we held our collective breath and exchanged ominous looks. "Here we go . . ." said Larry Grisolano, Pete's media consultant, stating the obvious. The attack we had been dreading was coming.

Except then it didn't. Rather than going after Pete, Harris trained her fire on Joe Biden, with a howitzer of an attack on his record of opposing busing. The Biden-Harris exchange dominated all the headlines coming out of the debate—the death blow we'd been preparing for never came. All in all, Pete put in a strong performance. He raised $1 million that night. And we lived to see another day.

Still, Pete paid a political price after the shooting. The media that had covered him uncritically up until that point turned on him. At first the coverage was negative. Then it was just . . . indifferent. He barely merited a mention in stories. I could almost hear the directives from editors. *He's had his fifteen minutes. It was fun, it was cute, it was different. Let's move on.* I was reminded of a quote from the April *New York* profile: "The problem is all of these candidates are going to have their moment in the sun. When the spotlight isn't on [Pete], and the world isn't moving organically for him, will it all still work?"

One good sign was that Pete's true-believer supporters never got discouraged. They were *hard-core*—among the most devoted of any candidate's. They created an intense online army that would welcome new Pete supporters with "digital hugs" and swarm any Pete detractors with fact checks. There were the diehards who dropped everything and drove to South Bend, where they'd stake out the Pete for America lobby, résumés in hand, until someone from the campaign met with them. There were the Barnstormers, a group of hundreds of supporters who traveled to Iowa, New Hampshire, and Nevada to campaign for Pete—oftentimes breaking out in dance mobs to Pete's campaign anthem "High Hopes." It got even weirder than that—on fan fiction sites like Archive of Our Own, there were chapters and chapters of Pete fanfic, many of them racy in nature. I'll spare you the details and give you one piece of advice: whatever you do, do not Google it.

Still—we needed a shot in the arm. Pete's polling had flatlined—he was no longer in the top three anywhere—nationally, in Iowa, or in New Hampshire. He'd fallen behind Sanders, Biden, Warren, and Harris in most polls. It's at times like these that campaigns show you what they're made of. A lot of campaigns and candidates sort of lose their footing—embracing tactics that reek of desperation (see: Beto O'Rourke announcing during a debate, "Hell yes, we're going to take your AR-15" or Kamala Harris staking her campaign on getting Donald Trump kicked off Twitter) and then enter a death spiral. We were determined to avoid those pitfalls, but we knew we needed to take some risks to get back on top.

First on my list? Reviving John McCain's Straight Talk Express. If the Straight Talk Express seemed ballsy in 1999, it was a downright crazy idea in 2019. While McCain's 2000 bus tour was 100 percent on the record, he didn't have to contend with social media back then—his bus didn't even have Wi-Fi! He could get away with off-color quips and policy discrepancies here and there because reporters weren't transcribing his every word and putting it all out on Twitter. We would have no such latitude. "Are you *sure* we should be doing this?" Pete asked me. We had no choice, I told him; we had to shake things up. And the bus tour would completely set him apart from his opponents, most of whom engaged only in highly controlled interactions with the media.

I took the challenge seriously, spending hours on the phone with former McCain advisers and former McCain beat reporters to plan out the tour. I asked them for their advice on best practices—everything from how many reporters we should let on the bus to what kind of food and drinks we should provide ("Booze, lots of booze" was a common refrain). It was a piece of advice I took: beer, spiked seltzer, wine (red or white), hard liquor—we had you covered. Our offerings fell short of the Straight Talk Express in only one regard: We didn't have a margarita machine on board. There's always next time, I guess.

We timed the inaugural launch of our bus to the Steak Fry, a uniquely Iowa event that was held in September and served as the unofficial kickoff to the caucuses. Campaigns prepared for the Steak Fry like it was the Super Bowl; it was a test of organizational strength—who could get the most bodies on the state fairgrounds outside Des Moines—and also of enthusiasm for the candidates. Widely covered in the national and local media, it was beyond critical for each campaign to put on its best show. Kamala had a marching band that accompanied her upon her entrance; Biden had a fire truck and hundreds of firefighters decked out in yellow (the firefighters union was one of his earliest endorsers); and Pete had the largest crowd of supporters—fifteen hundred in all—plus a big, beautiful blue-and-yellow bus that we affectionately referred to as the "Buttibus" or "Petemobile." (Sadly, we could never settle on a damn name for the thing—it was hard to top Straight Talk Express and my suggestion of Gay Talk Express never caught on.)

The spectacle lured in the press—larger print or TV outlets that might not have covered Pete's Iowa swing stacked the bus, eager for a front-row seat for the freak show. Pete and I were definitely nervous—this was relatively uncharted territory; how would it go? We quickly learned the answer: like a really, really awkward first date. After the reporters excitedly filed onto the bus and pulled out their cameras, iPhones, and notepads, they were like the dog that finally caught the car. They were so unaccustomed to having free rein with a presidential candidate that they froze up.

But after a couple legs, everyone loosened up a bit—the press got more comfortable asking questions and Pete got more comfortable giving answers. By the end, it was an unambiguous success. Cable news went live from the bus, major network morning shows like *Good Morning America* ran packages on it, and the local Iowa press—to whom we always gave the plum seats next to Pete—rewarded him with extensive, glowing coverage. Whatever we'd spent to commandeer the "Buttibus," we made back fifty times over in earned media coverage. More importantly, by the end of the bus tour, I realized something else: Pete had gotten his mojo back. Over the summer, during the doldrums of our campaign, he had adopted a workmanlike approach to campaigning—he was still out there giving speeches and doing events, but without much joy. The bus tour helped him regain the lightness and confidence that had catapulted him to contender status in the first place.

Now, we just had to build on it. Pete had coasted through the first three presidential debates, putting in strong, but not game-changing performances. The upcoming October 15 debate was the time to make a move and finally shift the momentum back to our side. The bus tour had helped, but the debate would be the big moment.

The polls showed him gaining in Iowa after months of stagnation. But there was a clear dynamic playing out in the primary that we needed to contend with. In the twenty-person field, individual Democratic candidates were staking out far-left positions—

positions way outside the mainstream—knowing that it would garner them attention and burnish their bona fides as "the most progressive candidate" in the race.

On the issue of health care, you had Sanders and Warren offering their unqualified support for Medicare for All (M4A), a slogan of a policy that was popular until people learned what it actually meant. The United States admittedly lags other developed nations in providing universal health care, but M4A was far to the left of what any other country had implemented—it would force every American into the same health plan and preclude the option of private insurance. The idea especially chafed workers whose unions had negotiated hard for their benefits and seniors who found the basic Medicare coverage lacking. Pete shared Sanders's and Warren's goal of universal health coverage—he just had a different view of how we could actually achieve it.

Larry, Pete's media consultant, impressed upon him the urgency of the moment: "Pete, you know what you believe. It's time to differentiate yourself. It's time to pick a lane." Pete had been loath to contrast himself with his opponents, but now it was time.

The debate started with a few perfunctory moderator questions about impeachment, but the focus soon shifted to the issue of Medicare for All. Marc Lacey, the national editor of the *New York Times*, asked Warren why she wouldn't come out, as Senator Sanders had, and acknowledge that Medicare for All would require middle-class tax hikes. She demurred, giving a wishy-washy answer.

Lacey turned to Pete: "Mayor Buttigieg, you say Senator Warren has been, quote, 'evasive' about how she's going to pay for Medicare for all. What's your response?" *Let's go.*

Pete fired up with a ferocity he hadn't yet shown on the debate stage: "Well, we heard it tonight, a yes or no question that didn't get a yes or no answer. Look, this is why people here in the Midwest are so frustrated with Washington in general and Capitol Hill in particular. Your signature, Senator, is to have a plan for everything. Except this."

To date, no one had laid a glove on Warren. Not only had Pete just challenged her on her policies, he'd gone after her core pitch: her competency and her message of "I have a plan for that." He went on that night to ably mix it up with other candidates like Beto O'Rourke and Tulsi Gabbard as well.

The debate was a legitimate "oh shit" moment. As smart as the press had found Pete, they'd always wondered, "Can he throw a punch? Can he be more than just the shiny, young object in the race?" Pete finally answered that question, leaving the debate with a new nickname coined by CNN's Van Jones: "Pistol Pete."

It didn't take long for our opponents to retaliate. Immediately, they started to unload their research files on Pete, prompting one TV reporter to confide in us: "I've never seen so much oppo pitched about one candidate." It was exhausting. From seven A.M. to ten P.M. every day, there was no end to the dirt that we had to push back on.

The behind-the-scenes animosity that Pete's opponents felt started to spill out in public. On November 9, the *New York Times* ran a story headlined: "Why Pete Buttigieg Annoys His Democratic Rivals." The piece walked through how Pete had united his opponents in their dislike for him: "a simple mention of Mr. Buttigieg's name . . . was enough to make Ms. Klobuchar extremely agitated"; Beto O'Rourke "was particularly aggrieved by Mr. Buttigieg, whom he viewed as 'a human weather vane' that represented the worst of politics"; "He got 9,000 votes in a college town that last voted for a Republican in 1964," [Governor Steve] Bullock added.

It wasn't exactly surprising to us that Pete's opponents were irked by him. I'd heard from cable news executives how candidates like Gillibrand and Klobuchar would call them up in the middle of the day after seeing Pete on air: "Why are *you* giving *him* so much time? *Some of us* are US Senators, he's *just* a small-town mayor."

I understood the resentment on a certain level; like most people, I've seen the movie *Mean Girls*. I just didn't get why these established politicians wanted to play the Regina George to Pete's Cady Heron. Each and every one of them had had a much bigger platform than Pete did when they got into this race. Maybe they should've been asking themselves and their staff: Why aren't I gaining in the polls?

The pace of the incoming fire seemed untenable. On November 16, the first Iowa poll in months was set to be released. None of us had the faintest clue what it would reveal—it had been months since we'd done any internal polling of Iowa.

The numbers were very closely guarded, but a reporter passed them along minutes before they were to be announced live, on air at CNN. I was stunned: the poll had Pete at 25 percent in Iowa, followed by Warren at 16 percent and Sanders and Biden tied at 15 percent. *What the living fuck!* Suddenly, the newfound attention from our opponents made sense—they were attacking Pete because he was winning.

The good news didn't end there. Three days later, St. Anselm College in New Hampshire dropped their poll of New Hampshire voters. It showed Pete at 25 percent, Biden and Warren tied at 15 percent. Sanders was in fourth place at 9 percent. The October debate had changed everything. *Everything.*

The narrative among the political chattering class shifted quickly from "Pete's fifteen minutes are up and his campaign is dead on arrival" to an even dumber one: "Is Pete peaking too early? Did he make too much of a move in that debate?" If Pete hadn't broken through in the October debate, he would have been completely written off by them. But now that he was dominating in November, it was somehow an indictment on his campaign and timing?

A presidential campaign is unpredictable. It's not like tracking a menstrual cycle and knowing the window when you'll be most fertile or the day you need to carry tampons in your purse. You've just gotta seize whatever opportunities come your way—whether it's "the time" or not.

TWELVE

Good Stuff

December and January were a slog. Pete was under attack every day. We dealt with faux outrage after faux outrage.

Warren started attacking Pete over his stint at McKinsey—the global consulting firm—during his early twenties. She demanded he release his client list from his time there, as if some twenty-four-year-old had been pulling the strings of global corporations. It was rich, too, because Warren had dragged her feet on releasing her tax returns from her time as a corporate defense lawyer.

Days before Pete was set to sit down with the *New York Times* editorial board, they piled on, running an editorial demanding that Pete be released from his nondisclosure agreement with McKinsey and make his client list public.

Pete faced a conundrum with no easy answer. He could either bow to public pressure and break his NDA, or abide by the terms of his NDA and continue to be attacked via insinuations from his opponents and their allies. The night the *New York Times* editorial posted, our campaign kitchen cabinet held a call with him. I led off by speaking for the group: "Pete, we're all in unanimous agreement here. We think you should break the NDA."

Pete pushed back hard: "I get the instinct and I appreciate everyone's advice. But I'm not doing it. This election is about character. I'm a person of my word. I signed my name on a legal agreement— I'm not going back on it, even if it hurts me in this election."

Other advisers chimed in: *Do you really want a legal agreement you signed in your twenties to tank your campaign? You didn't sign this knowing you'd run for president.* No matter what we threw at Pete, he was resolute: no way.

We settled on a middle ground. We'd put out a public statement from Pete calling on McKinsey to release him from his NDA, placing the ball in their court. Four days later, McKinsey announced that they would let him disclose his client list: "We recognize the unique circumstances presented by a presidential campaign. After receiving permission from the relevant clients, we have informed Mr. Buttigieg that he may disclose the identity of the clients he served while at McKinsey from 2007 to 2010."

When the news broke, I realized that our team had given bad advice in the heat of the moment. As much as we prided ourselves on being above the lefty social mob, we had capitulated. We had asked Pete to cross a line that was contrary to who he was, and that would have undermined one of his core strengths in an important phase of the campaign. As we would see during the 2020 general election, character does matter to the American people. I should've better understood the firm underpinnings of Pete's character before asking him to violate his personal code for a marginal political gain.

We were also contending with another dynamic in the race, one without precedent: how the media would cover Pete's historic candidacy as an out gay man. By 2020, reporters and politicos had wised up to the unfair constraints that women and candidates of color faced on the campaign trail. They jumped to Hillary Clinton's defense when she was called "shrill," to Obama's defense when he was accused of "palling around with terrorists." I wish I could say Pete received similar consideration.

To many reporters, who'd grown up in metropolitan areas, gone to Ivy League or top liberal arts colleges, and progressed to working at elite news institutions, there was nothing noteworthy or exotic about a national political candidate being gay. They had gay friends and gay colleagues. Marriage equality was the law of the land. So, case closed, right? Of course, they didn't see the hatemongers on the campaign trail calling him a "fag" and "sodomite" and flooding the campaign with death threats that required us to hire costly, round-the-clock security.

Meanwhile, the right wing had a field day, running with the stereotype of Pete—and gay men—as "weak" or "effete." The dog whistles weren't exactly subtle. In February, Rush Limbaugh—a common thread in my life, unfortunately—would attack Pete's viability in a general election, saying: "[Voters are] looking at Mayor Pete, 37-year-old gay guy, mayor of South Bend, loves to kiss his husband on the debate stage. And they're saying, 'OK, how's this going to look, 37-year-old gay guy kissing his husband on stage next to Mr. Man Donald Trump? What's going to happen there?"

Pete, to his credit, would respond to Rush at a Las Vegas town hall in a viral moment:

> *The idea of the likes of Rush Limbaugh or Donald Trump lecturing anybody on family values. I mean, I'm sorry, but one thing about my marriage is it's never involved me having to send hush money to a porn star after cheating on my spouse, with him or her. So, if they want to debate family values, let's debate family values, I'm ready.*

Pete could deflect these slights like a pro. But as a campaign, we were in a no-win situation. I'd learned about the restraint it took to fend off attacks like this when I worked for Obama. He never gave any race-based attacks against him even a hit of oxygen. There were a few things that were forbidden on the 2012 campaign; one of them was invoking the R-word—"Racist." We would attack birtherism, sure, but the calculation was that inflaming or elevating attacks of a racial nature would feed into this country's centuries-long prejudices against Black Americans.

We tried to do the same on Pete's campaign, but saw a backlash reminiscent of the one Obama experienced among segments of the Black community in 2008. He'd been called "not Black enough." Pete often found himself the target of attacks from the left that he wasn't "gay enough."

In the summer of 2019, the *New Republic* had run an outrageous piece from a freelance contributor labeling Pete "Mary Pete"—"The gay equivalent of 'Uncle Tom.'" The author went on to speculate:

> *The only thing that distinguishes the mayor of South Bend from all those other well-educated reasonably intelligent white dudes who wanna be president is what he does with his d*ck (and possibly his ass, although I get a definite top-by-default vibe from him, which is to say that I bet he thinks about getting f*cked but he's too uptight to do it).*

It was outrageous—in a liberal publication, no less. We never commented publicly on the piece, but we didn't need to. The backlash was immediate. The *New Republic* had been poised to host a September climate forum with all the Democratic candidates, but within a day of the column's publication, the League of Conservation Voters pulled out of it.

You would've thought that the *New Republic*'s misadventure

into the "not gay enough" genre would dissuade other news outlets from publishing offensive columns about Pete's sexuality. It didn't. BuzzFeed ran a sneering column titled: "You Wanted Same-Sex Marriage? Now You Have Pete Buttigieg," in which the columnist observed that Pete had "already fulfilled the gay assimilationist dream of marriage, the white picket fence, and a couple of rescue dogs." *Quelle horreur!*

The *New Yorker* would follow up with a column calling Pete "an old politician in a young man's body, a straight politician in a gay man's body." The author went on to chastise Pete's coming-out journey:

> *He didn't just wait until he was established in his political career. He also waited until after attitudes toward homosexuality had changed and same-sex marriage had become legal in more than half the states and was recognized by the federal government— all thanks to the courage and work of people who came out before Buttigieg did.*

Now, if the *New Yorker* writer had done the bare minimum of research, they would have learned that in Indiana, at the time that Pete was running for president, he could be legally fired from any job for the simple reason that he was gay. Unlikely for a mayor, sure, but a pretty significant sign of the culture.

I could generally roll with the punches with the media. It was one of the qualities that made me good at my job—I could remove emotion from my reactions to negative stories. This story line was an exception. I had seen Pete be called the worst homophobic slurs to his face. I'd seen him be told he was "going to hell." I'd witnessed a top network anchor ask him in an interview about his "lifestyle"—as if being gay was like choosing whether he'd have fish, steak, or salad for dinner. (The "lifestyle" question was cut from the segment that ultimately aired . . . by the anchor's gay producer.)

On the campaign trail, I witnessed the heavy interactions we would have with people struggling with their sexual orientation and gender identity. Teenagers who were queer or trans told Pete how they had been suicidal until they saw an openly gay presidential candidate. In a deep red area of Iowa, a seventysomething man approached Pete on the rope line, tears streaming down his face as he whispered to him his deeply held secret, one that he hadn't had the courage to tell anyone before. At an event in Key West, an older woman from Oklahoma waited in line to tell me how she'd left her community and family behind years ago because she'd feared she wouldn't be safe there if she tried living authentically as an out woman. After she finished the story, she introduced me to her wife.

Those are just a handful of the stories I heard. I can't imagine how many were shared with Pete. His candidacy was a validation for so many people in the LGBTQ community—people of all backgrounds—people who had been disowned by their parents, who felt relegated to the shadows before he'd run.

Did their experiences not matter? Of them, who qualified as "gay enough"? Did the New York–based authors of these self-righteous columns not understand that many in the country do not have the privileges they enjoy, and that maybe, maybe the visibility of an openly gay candidate was, in itself, a radical statement? Did they ever consider that someone deep in the closet might read their words about Pete being insufficiently gay and think, "If this is how I'll be received, what's the point of coming out?"

Perhaps the country wasn't ready for a gay president in 2020. But I believe Pete's candidacy spread more awareness about the lived experience of LGBTQ Americans. Specifically, that there is no right or wrong way to live your life—no matter what Rush Limbaugh or a *New Yorker* writer tells you.

During the December and January debates, Pete was the big target onstage. The never-ending attacks took a toll on his standing in the race. By late January, the polling averages of Iowa and New Hampshire put him in third place behind Sanders and Biden.

It was hard to put too much stock in the numbers—polling Iowa is basically impossible. The rules of the state's caucuses are arcane. Voters in Iowa don't just show up to a polling place when it's convenient for them, wait in line, and cast a vote behind a curtain. They have to show up at their local precinct at seven P.M. on caucus night, and publicly declare—in front of neighbors and friends—which candidate they're supporting. Once they do, they're cordoned off with the other supporters of their preferred candidate as the local election official computes the number of votes for each. Any candidate who doesn't get 15 percent of the assembled caucus-goers is immediately disqualified from putting any points on the board—their supporters can either cast their lot with another candidate or leave. So a candidate who gets 149 votes in a crowd of 1,000 is as irrelevant as one who gets 10. Then— this is where it gets really weird—the supporters of the candidates who fall above the 15 percent threshold have fifteen minutes to convince the people who caucused for the losers to join their side. When those fifteen minutes are up, the election official(s) tally the votes, declare a winner, and report it back. Are you exhausted and confused? Because after writing this, I am.

So, *how* do you poll Iowa? There was only one person who could—"The best pollster in politics" and "the polling Cassandra of Des Moines," Ann Selzer. She was a Des Moines–based pollster who'd accurately predicted every Iowa result since 1988 with her final pre-caucus poll. It was a testament to her dominance that no matter how many media or polling organizations tried to survey

Iowans, her poll was known as "*the* Iowa poll." In the small, inces-
tuous, and nerdy world of political polling, Selzer stood alone in
the one-name club—the equivalent of a Madonna or Elvis in pop
culture.

Two days before the Iowa caucuses, her poll was set to be released at eight P.M.—and not just in the *Des Moines Register*, as in
previous years. This time around, the *Register* was partnered up with
CNN, which planned to unveil the results live, on air from Iowa
with a highly-produced special anchored by Chris Cuomo.

That's why I lost it that morning when I got word from Ben
Halle—Pete's Iowa communications director: "We have a problem." By happenstance, one of our Iowa organizers had been on
the receiving end of a polling call from Selzer's organization the
day before. The call center employee went through the questions,
finally asking whom he supported. As she read off the names,
Pete's name was nowhere to be found. The organizer piped up,
telling her: "You didn't name the candidate I'm supporting: Pete
Buttigieg."

"Hold for a minute," she told him. "My computer might be
messing up." After several minutes, she returned to the call and
read the list again, this time mentioning Pete. She completely man-
gled the pronunciation of his last name, which is probably not all
that surprising, but you'd think someone who was helping to conduct the most important political poll in American politics could
get a tutorial in how to say: "BOOT-EDGE-EDGE." It's not rocket
science, but that wasn't my issue.

Ben told me how he'd reached out to Selzer to alert her to the
problem. She told him it must have been a one-time glitch—a
"font error." This was the most impactful and important poll in
presidential primary politics. It would set the narrative for the
caucuses, dominating the media coverage and dictating caucus
choices. I was beside myself. Hundreds of campaign staffers hadn't

been working months of ungodly hours to be at the whim of a poll conducted with all the thoughtfulness of a late-night drunk dial.

I called CNN immediately. And by that, I mean that I psycho-dialed CNN's political director until he picked up. After several missed calls on top of texts and emails, he hit me back: "What's going on?"

"So, there's an issue with the Iowa poll, do you know about it?"

"No?"

I walked him through my conversation with Halle and sent the contemporaneous notes our organizer had taken during the poll call: "Jesus, let me look and make some calls." It was a red flag that Selzer hadn't elevated the complaint to them. CNN didn't want to put its weight behind a problematic poll. As they dug into it, they couldn't get a straight answer from Selzer—whether it was a one-off font issue that hurt Pete or a system-wide error that left other candidates off the list that call center employees read to respondents.

This drama started to unfold as I was in the lobby of the Des Moines Marriott—*the* place to be in the run-up to Iowa caucuses, the Hawkeye State version of the *Vanity Fair* Oscar Party. As the clock counted down to the CNN poll reveal, the lobby filled with political luminaries such as Karl Rove, veteran columnists like Maureen Dowd, staffers from each of the presidential campaigns, and TV anchors from CNN, MSNBC, and Fox News. It was a veritable who's who of American politics.

I had no interest in the scene that day. I rode the escalator up and down from the lobby to the first floor as I pounded CNN with calls. At one point, I watched an intoxicated hotel guest tip over in slow motion and fall into the lobby fountain next to the escalator. I turned away as soon as I saw it—*Don't laugh. Don't laugh. I'm on a really important call. Please don't let me laugh.* It was a serious exercise in self-restraint as I saw hotel employees rush over and struggle to pull her out. Who knew, in a few hours, I could be doing the professional equivalent of falling face-first into a fountain if I didn't figure out this Iowa poll stuff.

Just over an hour before the live reveal, word started to leak out that there was a problem with the poll, and that Pete's campaign had raised it. The national reporters who had banked on doing little work that night, just filing an easy story off the poll, started to circle Mike Schmuhl, Chris Meagher, and me like vultures. They had to have this poll! How else would they have any clue what would happen on Monday?

We went into hiding in Mike's hotel room, and I issued an edict to them and our press staff back at headquarters—none of whom had the faintest clue what was happening: "Do not pick up your phone, do not respond to texts. Do not breathe a word of this. Just let me handle it."

Thirty minutes before the scheduled poll reveal, CNN abruptly changed their chyron (the text banner at the bottom of the screen on news channels) from "Iowa Poll Special" to "Iowa Caucus Coverage." Finally, when they went live at eight P.M. Central, they announced that they, in fact, would not be releasing the poll because of concerns with its methodology. I turned from the TV screen to Mike and Chris. "Guys! Guys! What the fuck, we just killed the Iowa poll!"

I never knew the actual numbers of the poll—they were never formally released or shared. Reporters with outlets that received the poll tweeted that it would've shown Sanders 22 percent—Warren 18 percent—Pete 16 percent—Biden 13 percent.

Those numbers would've killed Pete—his path to the nomination depended on a first- or close second-place finish. We were asking *so* much of Iowans—to take a chance on Pete, now a thirty-eight-year-old, openly gay mayor of a city of one hundred thousand people. As independent as caucus-goers are, no one wants to waste an evening standing in a corner for a guy who's gonna come in third place. It was a leap of faith to believe that he could be the guy to beat Donald Trump. They were almost there, but Iowans don't like to support third-place candidates, otherwise known as losers. Our big competitor in Iowa was Warren—she and Pete appealed to the same demographic—and that poll would've shaken our soft supporters and forced them to her side.

And there would have been serious ramifications for other candidates. The biggest loser, if the poll had been released, would have been Joe Biden. A two-term vice president, after all, was losing to a seventy-eight-year-old socialist and Pete.

Would it have stopped Biden from getting the nomination? I don't think we'll ever know the answer to that question. All I know is that—for the first time in its history—the Iowa poll was never revealed. Iowans would have to make their own choices, irrespective of the polls and media narrative.

As dramatic a moment as it seemed, the chaos was just beginning.

Without the Iowa poll to guide the coverage, reporters were largely rudderless going into the caucuses. Even among operatives on different campaigns, the general sense was that *anything* could happen. On our campaign, the consensus was that it was a contest between Pete, Sanders, and Warren. If you'd put a gun to our heads, we would have told you it was a 1, 2 fight between Pete and Sanders.

Election Day is its own special type of hell. If you've never worked on a campaign, you might think it would be the most hectic day of the campaign. In reality, it's the opposite. Months and months of working at breakneck speed culminate in a day where it feels like you're living in suspended animation. Every minute feels like an hour, every hour feels like a day, and a whole day feels like months of your life. Ninety percent of Election Days are reporters and anchors texting people like me: "What are you hearing," and people like me responding: "No, what are *you* hearing." All that stuff you hear from people on TV on Election Day is bullshit—nothing matters until the real, hard numbers come in.

Pete had started Election Day with his ritual—doing a road run with old friends from Harvard and South Bend. I started election night with an old friend—panic. Sitting in Pete's election-night suite with a reporter and a couple other staffers, the heaviness of the moment hit me. Some operatives might have basked in that moment. I ran to the bathroom and vomited my brains out from anxiety—the years of work weighing on me, the hundreds of staffers who had given up their lives to move all over the country for the campaign, and most of all Pete.

By the time the caucuses started, the suite had filled in a bit. Pete cut the tension he was feeling by throwing a foam football back and forth with a college buddy.

Less than an hour in, the networks started to release their entrance polls, and my anxiety started to subside. Pete was in either first or second place with *every* demographic. It was beyond anything we could have hoped for. The data was bolstered by the numbers that our organizers were reporting as they returned from the field.

It became increasingly clear that it was going to be a photo finish between Pete and Bernie. And then disaster struck—the Iowa Democratic Party announced that there had been a glitch with their caucus night app, which precinct captains used to record results at each caucus site, and thus they would not be able to call a winner that night. We had drafted two speeches—one for a victory and one for a disappointing result. As the clock approached midnight, we knew that Pete had to go out and make a speech. He had to say *something*, and he had to make his flight to New Hampshire, where he had a full day of events scheduled.

As a group, we made a decision: "Fuck it." By every objective measure, this was a shock result for Pete. A first-place or close second-place finish for him was a "win" in anyone's book. He went out to a screaming crowd and gave a rousing speech in which he didn't outright declare victory but announced that his campaign was moving on to New Hampshire "victorious."

That soundbite, plus the visuals of our event—throngs of fired-up Pete supporters waving their signs—earned Pete more TV coverage than his competitors. And it helped shape the narrative going into the next day that Pete was *a* winner, if not *the* winner of the caucuses.

We got some blowback from our opponents and critics in the media that we'd jumped the gun. Sanders, who'd previously swatted away Pete like a gnat, was especially incensed. Though I suspect it was largely because we'd outmaneuvered his team and done the smart thing. None of it mattered when—the next day—the Iowa Democratic Party released their first canvass of the caucus results. Pete was in first place. "How about that?!" he responded in his characteristically Leave It to Beaver, earnest way when Chasten and I told him that he had just won the first presidential contest and earned a place in the history books.

Pete's event in Laconia, New Hampshire, that evening started a little bit late, but when he went out to address the electrified crowd, all the networks carried him live as he declared victory (sort of) for a second time. He choked up with emotion as he reflected on the barrier-breaking nature of his win—the first-ever for an LGBTQ presidential candidate: "This validates for a kid somewhere in a community wondering if he belongs or she belongs or they belong in their own family, that if you believe in yourself and your country, there is a lot backing up the belief."

Still, it was hard not to feel like he had been robbed. He should have been the big story coming out of Iowa. In any other election year, his win would've dominated national and even global head-lines. Instead, they were focused on the Democratic Party's inability to conduct a caucus. It also cost us handsomely when it came to fundraising. The $2 million we raised in the twenty-four hours after Iowa was the best fundraising period of our campaign, but if he'd been declared an outright winner that night, our finance director predicted that he could've raised upward of $20 million.

For all the talk of Pete's charmed political existence, he was probably the unluckiest candidate in the race.

We trudged our way through the next few weeks. Pete came in a close second to Sanders in New Hampshire—the closest 1, 2 finish in New Hampshire primary history. Heading into the Nevada caucuses, we could feel our momentum fading. There was the Iowa clusterfuck; Sanders's close win in New Hampshire over Pete deprived us of another key "bounce"; and then there was the fact that former New York mayor Mike Bloomberg was flooding the fifteen Super Tuesday states—the states that would likely crown the next nominee—with over $215 million in TV ad spending—completely crowding out campaigns like ours. To put that spending in context, it was more than twice what Pete raised for the *entire* 2020 campaign cycle.

Still that week in Nevada wasn't all bad—we were in Vegas, after all. Our campaign staff spent mornings and afternoons prepping Pete for the Nevada debate, late nights playing blackjack, and one memorable evening watching a billionaire self-immolate onstage. By mid-February, there wasn't much love between any of the Democratic candidates, but there was one feeling we all shared: *Fuck Mike Bloomberg.*

Every Democratic candidate had spent the past year campaigning, visiting living rooms in New Hampshire and Iowa, hustling for every vote. Bloomberg refused to participate in the early contests, knowing that they would be decided on retail skills and message—not money. He was thumbing his nose at the other candidates, as if he was above it all, and was skating through the process with little scrutiny, hoping to carpet bomb the airwaves and ride to victory. It was all going according to plan until the Nevada debate.

Pete, along with candidates like Sanders, got his shots in at Bloomberg. But to her credit, Elizabeth Warren was the one who threw the fateful grenade:

"So I'd like to talk about who we're running against, a billionaire who calls women 'fat broads' and 'horse-faced lesbians.' And, no, I'm not talking about Donald Trump. I'm talking about Mayor Bloomberg."

Bloomberg was like a deer in headlights—for all the money he'd spent trying to buy the election, it appeared that he hadn't spent a dime on debate prep or, even worse, a moment thinking about why he should be president. He was universally declared the loser of the debate, and his support immediately began to recede.

I'd always thought of Las Vegas as the city of lights—a place that, no matter the time of day, was permanently neon and buzzing with life. Caucus day was uncharacteristically gray, cold, and rainy. The strip seemed empty, and the world's largest Ferris wheel sat motionless, suspended in the storm. Pete stared at it silently from the window of the suite where we were watching the results come in, biting his cuticles and retreating into himself.

It had been just nineteen days since he'd won the Iowa caucuses and made history—nineteen days since he'd been at the top of the Ferris wheel of presidential politics, with all the bright lights below him. Now those lights were dimming as the Nevada caucus results came in, placing him in a distant third behind Sanders and Biden.

No one in the room voiced the obvious: it was the beginning of the end. I still remember the pundits on TV that night saying:

"Where does Mayor Pete go from here?" It was the same question they were asking of Elizabeth Warren and Amy Klobuchar, sure, but they hadn't won any contests. They hadn't felt that unique thrill of victory.

Five days later I found myself sitting on the floor in the corner of the mostly empty lobby of the Hilton Garden Inn in Charleston, South Carolina. It was the only perch I could find that afforded me access to both a phone outlet and the hotel bar. I was on a call with Pete's kitchen cabinet. We were discussing next steps.

I wasn't blind to the realities of the campaign. But it didn't make them any less painful as I listened to three out of the five advisers on the call broach the possibility that Pete should drop out of the race if he placed third or fourth in South Carolina, as was expected. They weren't wrong. Not at all. But it was hard to listen to them discuss it so clinically.

I unmuted my phone and told them under no certain terms would we raise this with Pete before the South Carolina vote came in. "It will kill him, especially after how far he's come. There is *no* talk of dropping out before Saturday."

I was protecting myself as much, if not more, than I was protecting Pete. When I muted my phone once more, I found myself hiding behind the veil of my hair, surreptitiously wiping away tears. Elizabeth Warren and her staff were staying in the same hotel, and the last thing I wanted was for them to leak to Playbook that a Buttigieg aide was spotted crying on the floor of the hotel lobby on the eve of the primary.

Leaving South Carolina was rough. It was the afternoon of their primary and no votes had been counted, but there was little reason for optimism. We flew to a late-afternoon rally in Nashville, Tennessee, where Pete attracted an impressive crowd of over five thousand people. From there, we flew to an absolutely packed arena in Raleigh, North Carolina, where Pete delivered a speech as the South Carolina results came in. Our final flight of the night was to a small airport outside Plains, Georgia. I'd helped set up a meeting between Pete and former president Jimmy Carter for the next day—a meeting that we had initially intended to signal that we were in this race to stay.

When we landed, Pete and Chasten got in the lead car. I hopped in the one behind with other staffers. We had a call scheduled with Pete and our senior advisers, and it was brutal. After our senior strategist, Michael Halle, walked through the results that had come in from South Carolina, Pete jumped in as nonchalantly as he could: "Gang. Unless I'm missing something here, the next step seems pretty obvious." He was going to drop out. The conversation immediately shifted to planning a concession in South Bend the next evening.

I kept a fake smile plastered on my face the next morning. There was the remote *Meet the Press* hit, riddled with technical difficulties, where Pete gamely answered Chuck Todd's questions, giving no indication of where his head actually was that day. Next up was Pete's breakfast with Jimmy and Rosalynn Carter. I sat outside in a volunteer's minivan for the hour, using the time to make phone calls and put on some makeup.

When Pete emerged from their house, he beckoned me over to walk with him to the café where the press was waiting for him. The Carters made the last-minute decision to join him for the event, but they would be driven over with Chasten. It wasn't a long walk—maybe four city avenues—but it was just him, me, our campaign photographer, and security as we started the stroll. It was as intimate a moment as you can get on a presidential campaign.

I asked him how his breakfast with the Carters had gone. He told me that Carter had been matter-of-fact from the beginning:

"You've had an amazing run, but you know it's over now." We paused from time to time to marvel at the Carter landmarks and snap photos—even with everything that was going on, we couldn't help ourselves. When we stopped at the Plains railroad tracks, I choked up as I asked Pete if he'd ever imagined—way back when we first met—that we'd find ourselves here in Plains, Georgia, after he'd won Iowa and come a close second in New Hampshire. He admitted that he hadn't, but that he had no regrets. Then he thanked me for believing in him.

The scene at the Plains café was surreal. Local community members packed the venue along with our traveling national press corps. The press lost their shit when the Carters walked in—it wasn't every day that you got to be in such close proximity to a former president. As Pete took a seat at the table with them, Carter greeted the press and asked them each to name their outlet. He was very much in Sunday school mode.

He was also deviating from the plan. Like any former US president, he had a tightly knit group of former aides advising him—aides who the day before had told me that there was no way that the Carters, if they somehow decided to join Pete at the event, would want to talk to the press. But what was I going to do? Tackle a ninetysomething former president?

Until Carter had opened up the dialogue with the reporters, I don't think any of them had planned on asking him questions. There is a certain amount of formality and respect that still exists when you're in the very rare presence of an elderly former president. But now all bets were off. When one of the reporters asked Carter what was next for Pete, he responded: "He doesn't know what he's going to do after South Carolina." *Gulp.* He wasn't wrong. But we didn't need the press to know that. After a couple more questions, I shooed the press away. Thankfully, they didn't read too much into Carter's comments.

We had a couple of hours in Plains between the Carter event and our flight to the next destination on our itinerary: Selma, Alabama, where Pete would be joining the other presidential candidates for the annual march with John Lewis across the Edmund Pettus Bridge. Lewis was in declining health, and he was unlikely to march, but it was an event imbued with historical importance. After all, it was the site where in 1965 Lewis had led a peaceful protest across the bridge, only to find himself and his fellow marchers viciously beaten by state troopers and police officers in an incident later dubbed Bloody Sunday.

We had set aside an hour of press call-ins to Super Tuesday states, and an hour for downtime, which was spent instead on the phone with Pete's advisers. We ran through the details for the secret South Bend event that night; he wanted to make sure it went perfectly.

Then we turned to the thornier issue: whether he should endorse another candidate. "*This* might be ending tonight, but I still want to have an impact on this race. And I don't want to rush into anything." There was little disagreement about where his endorsement should go if he offered it. Biden was the closest to him ideologically of the remaining candidates, and he would be the strongest general election candidate against Trump.

Analytical as always, Pete asked each of us to make the case for it. He went around the call from one staffer to the next, lobbing questions at us like it was a routine call: How would Democratic voters view an endorsement? Would it inflame divisions among the electorate or help heal them? What would the press reception be? Would it actually make a difference? The last question was the sticking point for him.

Satisfied with our answers, he told us that he was inclined to endorse Biden, but that he wanted to sleep on it and wait until the next day. Before we hung up, he asked optimistically, "Soooooo,

Lis, does this mean my press calls for the next hour are off?" They weren't, I told him; he'd have to end the campaign as he started it—talking to everyone. If we canceled the press calls, there was a chance that people would start putting two and two together. We did not need the news to leak out early. As always, he was a good soldier.

It was a stiflingly hot and humid day as the presidential candidates lined up with civil rights leaders like Jesse Jackson and Al Sharpton to re-create the 1965 trek across the bridge. Bloomberg showed up late with a coterie of his hired former NYPD bodyguards and tried to cut the line. He was unsuccessful. Thankfully, money can get you only so far.

I tried to keep my eye on Pete, but in the waves and jostling of the crowd, I lost sight of him a half mile before the bridge. The last glimpse I got of him, I saw tears rolling down his face. Maybe it was the gravity of the moment, being present for what was likely to be John Lewis's last walk across the Edmund Pettus Bridge; maybe it was reality setting in, knowing that he was about to end his presidential campaign.

I cut through every narrow opening in the crowd, ducking under raised arms and shimmying between couples. The march was so packed that cell service was shot—none of my calls or texts were going through.

I started to panic. In that moment, I felt him slipping away. *Is this how this campaign ends for me?* I envisioned the ever-punctual Pete facing the press on the plane alone and flying back to South Bend as I tried to navigate the crowded streets of Selma. Finally, I found another staffer, and we commandeered a ride back to the airport.

Before we let the press on the chartered plane, Pete and I sat in the back, talking through the brief remarks he'd need to deliver to the press before takeoff. We'd save the big reveal for South Bend, but at the very least, he had to share that the plane would be headed to South Bend and not Texas, as planned. "Should I take any questions afterward?" he asked. I told him to take one question, max: "You've answered all of their questions—they can quote your speech in South Bend."

After the press boarded the plane and settled in, I told them to get their cameras and recorders out. I forced a smile as I asked Pete to come out from behind the curtain that separated him from the press. He looked downright tired. I noticed he was swallowing hard between sentences—an effort to keep his voice steady and emotion at bay:

"Um. A little bit of news for you about our flight. We're, uh, making a change in our travel plans and traveling to South Bend rather than to Texas. We're, uh, gonna be making an announcement there about the future of the campaign. And we, uh, uh, look forward to sharing with our supporters and the country where we're going from here. And so that is why, uh, you will find us headed in a different direction than we'd initially planned on."

The uhs and ums were a tell; Pete always spoke in complete, perfectly formed paragraphs.

A CNN reporter pounced. "Are you getting out of the campaign?"

Pete was smiling, but blinked twice before answering, his eyes beginning to glisten: "I look forward to making an announcement tonight to the country." Then he walked back to the private area that he shared with staff. I stayed in the cabin with the press to enjoy the last moment I'd get with them, watching as they scrambled to call their editors and file the news. Scramble they did, but it was notable that none of them yelled questions back at Pete or tried to corner me with questions. We'd always been straightforward with them. In those very painful seconds, they extended us a much-needed moment of grace.

When I joined Pete behind the curtain, I found him with his

iPad, working furiously on his concession speech, a bottle of beer in the beverage holder next to him. I joined him in drinking a beer and making edits to the shared Google Doc that contained the final remarks he'd give as a presidential candidate. As usual, we worked in tandem: I could see his edits in real time; he could see mine. The only difference was that I was making mine while fat, hot tears rolled down my face.

As I stood over Pete's shoulder, pointing out the edits I'd made and my justification for making them, a few of my tears fell onto his iPad. "God, this is so embarrassing," I apologized. "I just can't help it."

"Don't worry about it," he responded. "This is good stuff."

People, smart people, often say that work isn't life, that your work *can't* be your life. And they're right. I'd done a piss-poor job of nailing that balance.

Living in New York opened my eyes to a world that didn't revolve around politics—my friends didn't work in the industry, they couldn't pick Mark Meadows or Kevin McCarthy out of a lineup, and if they took the time to vote, it wasn't necessarily for a Democrat.

Work *shouldn't* be your life, but what Pete's campaign taught me is that it could *change* your life. I'm not going to lie and say that I had a vibrant life outside of that campaign—it was an all-consuming project.

As much as I pretended the tabloid stuff and my experience with Eliot hadn't affected me, it had, deeply. It closed me off to being able to trust people on a personal level or enjoy deep intimate relationships.

I wasn't some lamb before I'd met Eliot. I never avoided controversy for the sake of avoiding controversy. I certainly never passed up an opportunity for a smart-ass quip or condescending comment. I *thought* I knew who I was. But unconsciously, I started to become someone else, the tabloid version of me.

It had taken thirty-seven years of my life and working with Pete to realize something for the first time: I, a flawed person, was enough as I was. I didn't need to pretend to be anything or anyone else.

How I looked, talked, acted, thought about the outside world and what they thought of me didn't matter. I *could* be weird. I *could* be shy. I *could* be unnecessarily brash. I *could* be overly tough. I *could* be incredibly vulnerable. I *could* be a cynical tactician. I *could* be idealistic and have a long view of politics and Pete's place in it. I *could* be more than a caricature. I *could* be something that I'd never imagined myself to be.

Pete had a line that he used in interviews: "So much of politics is about people's relationships with themselves. You do better if you make people feel secure in who they are."

He was right. He saw me for who I *actually* was and, for the first time in my adult life, I did, too.

THIRTEEN

The Prince of Darkness

By any measure, Pete's campaign had been a raging success. He'd established himself as one of the brightest stars in the Democratic Party, became the Biden team's go-to TV communicator, and then was appointed Transportation Secretary—a crucial Cabinet post given how infrastructure was at the forefront of the Biden administration's agenda.

You might be asking yourself then why, a year to the day after Pete's exit from the presidential race, I found myself on a call not with Prince Charming but with the "Prince of Darkness," the political insiders' nickname for New York governor Andrew Cuomo, who was facing allegations of sexual harassment.

Well, it was complicated.

It started with a cold call from him in the spring of 2018. His right-hand woman, Melissa de Rosa, gave me a two-minute heads-up before the "No Caller ID" popped onto my iPhone screen. *Ugh*, I thought. It was a Monday night, and I was inundated with work. I needed a call out of the blue from the governor of New York like I needed an invitation to a one-year-old's birthday party.

So. Some back story here. I didn't leave New York politics after 2013. I worked on races at every level, including for people like Adriano Espaillat, the first Dominican elected to Congress, and Eric Gonzalez, a reform-oriented prosecutor who became Brooklyn DA. Still, in all that time, there was one New York politician I had never met: Andrew Cuomo.

My view of him was shaped largely by his press—politically effective, ethically iffy (he'd been the subject of a federal corruption investigation), and a total asshole. Then there was what Eliot had told me about him. Essentially, that Cuomo was Rosemary's baby several decades down the line.

To say Eliot hated Cuomo would be an understatement. There are a lot of things I hate in life: genocide, racism and sexism, child and animal abuse, bats (the flying rodents, not Louisville Sluggers), chewing gum, and small talk, to name a few. But never have I ever loathed anything or anyone as much as Eliot hated Cuomo. Eliot had convinced himself that Cuomo was the catalyst for his political demise. It all went back to a report that Cuomo's attorney general's office issued a year into Eliot's first term as governor, alleging that Eliot abused his access to the state police and used them to spy on his enemies. Eliot's animus reminded me of a Herman Hesse quote I'd read in high school: "If you hate a person, you hate something in him that is part of yourself. What isn't part of ourselves doesn't disturb us."

The Andrew Cuomo I met over the phone that night was far from the devil's spawn I was expecting. He was charming, self-deprecating, and thoughtful. We spent over an hour talking about politics and governance—what they meant to each of us. We talked about our dads. His father, former governor Mario Cuomo, had passed away a couple years earlier and mine was in increasingly poor health. As the conversation wound down, he asked, "So, you're a yes?" I told him I was a maybe. I knew working for him was a risky proposition, but—for good or for bad—risk has always been sort of my thing. I decided this one was worth it. Within three weeks I was on his payroll.

Not that he really needed me in that race. Even though media and political insiders took issue with Cuomo's Raging Bull persona and tactics, it was hard to argue with the results he'd yielded as governor. In two terms, he'd racked up a list of legislative accomplishments that included marriage equality, tough gun-control laws, a $15 minimum wage, paid family leave, and tuition-free college for working—and middle-class—New Yorkers. "You cannot spell the word 'progressive' without progress," he'd tell editorial boards and his left-wing critics. The people of New York agreed. He easily defeated *Sex in the City* actress Cynthia Nixon in the Democratic primary and sailed through the general election. Once he won a third term, I was off his campaign and off to the races planning Pete's campaign for president.

We continued to keep in touch as he rode high as "America's governor" during the COVID-19 crisis, his daily briefings broadcast live on national TV. "So, Lis-beth," he'd ask, using his preferred nickname for me, "What do you think? How am I doing?" He rarely asked questions that he didn't know the answer to, and the answer was evident from the fawning media coverage he was receiving: well, exceptionally well. In those days, you couldn't turn on the news, log in to social media, or even walk down the street without seeing or hearing people falling all over themselves to praise him. For once, it felt like the larger world was getting to see the Cuomo I had come to know: relentless and good at bending people and government to his will, sure. But also deeply empathic and human.

The hype got out of control, as hype so often does. The bougie,

Resistance-friendly Lingua Franca brand sold $400 cashmere sweaters hand-stitched with CUOMOSEXUAL across the chest. (*Cuomosexual* was the cheeky, cringeworthy term that had been coined for loyal watchers of his briefings.) He won an Emmy Award for his COVID-19 briefings. He was floated as a replacement for Biden on the 2020 ticket (delusional) and coronated as a front-runner for the 2024 Democratic nomination (fever dream).

He started to feel his oats. Just four months into the pandemic, he signed a multimillion-dollar book deal with Random House to tout his leadership lessons during the pandemic. It was the height of hubris. It was as if the head coach of the Atlanta Falcons had walked off the field during the third quarter of the 2017 Super Bowl, satisfied enough with his team's 28–3 lead over the Patriots to write a book about lessons in winning the Lombardi Trophy. As everyone knows by now, Bill Belichick and Tom Brady overcame the biggest deficit in Super Bowl history to win that game 34–28. Just as the Patriots came roaring back, COVID would, too, with a second devastating wave that took the lives of fourteen thousand more New Yorkers—all while Cuomo was promoting his memoir.

He also began to dig in his heels about decisions he'd made early on in the pandemic, specifically the policy of returning COVID-19 patients from hospitals to their nursing homes. New York was hardly alone in implementing that policy, which stipulated that the patients had to be medically stable and that the facilities had to be able to properly care for them. Other states like New Jersey, Pennsylvania, and Michigan followed the same guidance. Whether the policy was medically sound or not, Cuomo allowed the controversy around it to snowball. His administration was accused of undercounting the number of New York State COVID-19 deaths in nursing homes, attributing some of them to hospitals instead. It was a serious charge—and one that the Cuomo administration vigorously disputed, noting that they assigned deaths to where they occurred, not to where the deceased had contracted COVID. But they were slow to release the supporting data, which didn't help their cause. At best, it was interpreted as incompetence. At worst it was seen as a cover-up.

Simply put, the nursing home scandal was a self-inflicted PR wound born out of the governor's stubbornness. If Cuomo had simply come out and said that his administration was following the science the best they could, that there was no playbook for handling a global pandemic, and that they had made some missteps along the way, people would've understood. But he refused to show any humility, and many in the media assumed his administration was juking the stats to make him look better. The crisis reached a nadir at a January press briefing during which he was confronted about the undercount. Visibly bristling, he defiantly declared: "Look, whether a person died in a hospital or died in a nursing home, the people died . . . Who cares?" "Who cares" are two words that should never come out of a politician's mouth. *Especially* when it has to do with people dying.

America's governor was quickly turning into America's asshole. And like most assholes, he'd soon get a wakeup call. Except in his case, it was more like an air raid siren.

On February 24, 2021, a former administration employee accused him of sexual harassment. Within hours, the allegations were picked up far and wide in the media. And once again, I was getting roped in with a small group of outside advisers to help Cuomo navigate the crisis. The ask was innocuous—"just a few calls." The assumption was that this would be a one-day story. *Famous last words.*

My decision to say "yes" was grounded in the fact that I believed Cuomo's denial of the allegations, which had seemingly come out of left field. He'd been a champion of the #MeToo movement—and in those days, I'd never heard so much as a whisper about his personal conduct. Could he be flirtatious at times? Yes. Did he occasionally make jokes of a sexual nature with staffers—both male and female—at the workplace? Yes. Was he unusually into physicality as a modern-day politician? Also yes. He fashioned himself after leaders like LBJ, who was known for grabbing a lapel or twenty in his day.

The easier answer, naturally, would have been no. But politics is filled with cut-and-run artists—soulless social climbers who cling to elected officials when they're popular, then disappear the second they're not. I never wanted to be one of those people. I still struggled with the emotional scars of my own brush with scandal. Years of therapy and professional rehabilitation could never undo the devastation I felt when some of my friends and colleagues turned their backs on me. It's impossible to describe the isolation and hopelessness and darkness that consumes you when you're in the eye of a PR shitstorm. I couldn't live with myself if I let anyone I knew well go through it alone. Obviously, my predicament was pretty different—I wasn't accused of doing harm or wrong to anyone else, but in my eyes—at the time—there was little distinction.

From the beginning, I had reservations. Advising Andrew Cuomo on communications is notoriously an exercise in masochism. I'd largely escaped it on his 2018 campaign, where I'd worked mostly from home and his headquarters in Manhattan on bigger picture campaign issues. But—from what I heard—I was an exception. As successful and powerful as he was, he somehow woke up every morning with the worst, darkest, and most twisted ideas in his head. He'd handwrite, dictate, or BlackBerry pin (never email) to underlings what was on his mind, oftentimes before six A.M. Hours of his aides' time every day and week would be wasted talking him down from his harebrained schemes.

Faced with the first accusation of sexual harassment, Cuomo

swore to the crowd advising him that nothing, *nothing* else would come out. It didn't take long for us to see that he wasn't being completely truthful with us. And quite possibly with himself.

Just a few days later, a twenty-five-year-old former executive assistant in the governor's office came forward with more accusations in the *New York Times*. She recounted how the governor had repeatedly inquired about her relationship status, talked about how he was lonely, and tasked her with finding him a girlfriend—actions that she interpreted as sexual advances.

Behind the scenes, Cuomo conceded that he had been "stupid" to engage in any personal conversations with a female staffer that he barely knew: "I should have said, 'this is fucking trouble.'" Still, he denied any malintent behind any of his comments.

Like others on the team, I began to feel a sense of unease—the allegations were at the very least *creepy* and they showed extremely poor judgment. We began to have side conversations where we vented about our discomfort with the situation. Aides from his early days as attorney general and governor were especially dismayed. One told me, "I'm in disbelief. He used to have a rule about never being alone with a woman in his office under any circumstances. Now he's having these sorts of conversations with a twenty-five-year-old? What the hell is going on up there?"

We had been told there would be no additional allegations, but here was this one—above-the-fold on the front page of the *Times*. It didn't feel like we were getting the whole truth. And that was a big problem, not least because the number one rule of crisis communications, the most *sacred* rule of crisis communications, is that the person in crisis needs to be completely truthful with the people advising them. To give good advice, we had to know what other potential stories were out there. Then, of course, there was the fact that so many of us were putting our reputations on the line for him. This wasn't just about him—he was asking us to take all of our time and collective talent to advise him and to risk our names with reporters and elected officials

Fool me once, shame on you.

We swallowed our doubts and tried to help Cuomo weather the storm.

Politicians had survived worse allegations of sexual misconduct—most notably, Presidents Bill Clinton and Donald Trump. On the other hand, there was Senator Al Franken, who stepped down from office in 2017 after several women came forward and accused him of unwanted touching and kissing—accusations that Franken vigorously disputed. But Democrats—including some of Franken's colleagues who had called for his resignation—had regrets about how it all went down, questioning whether he'd received adequate due process. The complicated politics of the #MeToo debate reached a fever pitch with the media circus around Supreme Court Justice Brett Kavanaugh's Senate hearings, which Democrats like Claire McCaskill blamed—in part—for their losses in the 2018 midterms. The hearings had spurred a backlash with voters who believed that #MeToo was being unfairly weaponized for political ends.

In our internal conversations, we talked about the lessons from the Franken and Kavanaugh controversies and what we could learn from them. However, the case study we looked to the most had nothing to do with sexual harassment or misconduct: it was the curious case of Governor Ralph Northam in Virginia. In February 2019, a right-wing blog published a photo from Northam's 1984 medical school yearbook, which they alleged showed Northam hamming it up in blackface next to a classmate wearing a Ku Klux Klan costume. Within forty-eight hours of the photo being posted, everyone from the Virginia Democratic Party to former Virginia Governor Doug Wilder—the first Black governor ever elected in the country—Democratic presidential candidates like Joe Biden and Kamala Harris, and the *Washington Post* editorial board called on Northam to resign. Northam's handling of the incident didn't inspire a ton of confidence—first, he apologized, then he denied he was in the photo at all—even as he admitted that he'd once worn blackface when he'd competed in a Michael Jackson dance contest. He refused to leave office.

The political world scoffed at him. CNN ran an online column from its top political analyst Chris Cillizza confidently declaring: "Ralph Northam has to resign, even if he doesn't know it yet." But then a bizarre thing happened—poll after poll showed that while Virginia voters as a whole were evenly split on whether or not Northam should step down, Black voters—the constituents that political prognosticators were certain would be the most offended by the photo—believed, by a large margin, that he should stay in office. And he did. A year later, his job approval rating soared to 60 percent among all voters.

The decision was made. Cuomo would "Northam it." It was a ballsy strategy, sure, but it was the only hand he had to play. He called for due process and authorized the New York Attorney General's office to conduct an independent investigation into the sexual harassment allegations. He held a press conference to make his case directly to the people of New York—one that was carried live by local TV networks across the state and every national cable news network.

We prepped for the press conference at the Albany governor's mansion. A group of ten of us hunkered down in the poolhouse behind the main house—feet away from the shallow hot tubs where Franklin Delano Roosevelt, as governor, had exercised his polio-stricken legs.

Everyone was on edge and exhausted. There was one main exception: the governor. He showed up to the prep session as cocky, casual, and self-assured as ever. He made small talk and cracked jokes. Outside of the seemingly never-ending stream of Nicorette that he popped into his mouth, jaw tensed, you'd never have known that he was under any sort of stress. He was in a fight for his political life, but he was the only person who didn't seem to know it.

What, me worry? It was as if after nearly eleven years in office, he felt untouchable. He'd lived through federal investigations, a top staffer going to prison, and a global pandemic that took the lives of tens of thousands of New Yorkers. What were a couple allegations of misbehavior?

I led the prep, giving me the opportunity to look him in the eyes as I peppered him with questions about his conduct. "Have you ever acted inappropriately toward women in the workplace?" *No.* "Have you ever had inappropriate relationships with women on your staff?" *No.* "Do you think other women will come forward?" *No.* Other advisers jumped in with questions and received the same forceful feedback. There's no way he would just lie to *all* of our faces, we concluded. *What* kind of person would do *that*?

We all began to wonder what kind of person we were working for a week later when we got word of new allegations—the most serious and shocking yet. The *Times Union* was working on a story about how a current employee of the governor's office had hired a lawyer and was claiming that the governor had groped her at the Executive Mansion.

What. The. Fuck. That's the only way to explain the reaction among the advisers, especially the women. It started to feel like we were being manipulated—used because of our gender to cover and lie for him.

"This is disgusting, right?" I asked his former communications director, who was also advising him from afar. "Did you see any of this?"

"No, it's *so* disgusting," she told me. "I don't even know what this is."

Another adviser was even more direct in a call with me after we heard the latest allegations: "He is *dead. Dead.* We just need to figure out how to land this plane."

It was tempting to cut the cord right then and there, but instead we waited until we heard directly from Cuomo himself. Again, everything followed a similar rhythm. Within a couple of hours, Cuomo was on the phone with us vehemently denying the allegations. There was one key difference—one that took me a minute or two to process. I heard something I'd never heard in the governor's voice before—*fear*. Genuine *fear*. "This is *not* true. It *never* happened," he told us.

In real time, we could hear the most powerful person in the state of New York beginning to process that he was in real trouble—not one-day- , one-week- , one-month-bad trouble, but resign-from-office trouble. Unlike on past calls where he deferred to staff and advisers, he wanted to come out guns a-blazing against the accusations. He didn't have time for reason, he was all emotion. "Bad Andrew"—as staff privately called him when he got into his worst and darkest moods—was making a comeback.

He wanted to accuse his accuser of having financial motivations. He wanted to expose her for hiring a notorious Albany-area ambulance chaser. He wanted to go after her character head-on: "If I don't fight back, why don't I just resign?" It took the force of everyone on the call to talk him off the ledge and convince him how disastrous it would be to go that route. We pleaded with him to show some humility and contrition. The person who finally got him to back off was an unlikely participant on our calls: his brother, the CNN anchor Chris Cuomo. Chris was oftentimes the last bulwark against his brother's worst instincts.

While Chris could sometimes be a dick to staff and informal advisers, reminding us: "I work in the media, you don't" or "I know this business, you don't," he was far from the goon he was portrayed to be in the media coverage that ultimately led to his firing. He could be more direct with Andrew than any of us could be. He leveled with him on calls, telling him—in no uncertain terms—that his behavior was inappropriate, that he needed to be more apologetic, and that he could never, ever come across like he was attacking his accusers. If we didn't wrap up a call with a resolution, Chris would usually end it with: "Andrew, pick up your phone. I'm calling you after we hang up." And he'd get his brother to agree to the direction laid out by the cooler heads around him.

While I appreciated Chris's loyalty to his brother and his commitment to getting him to do the right thing, I didn't fully understand the dynamic between them. After I asked about it once, one of Andrew's longtime advisers told me how he felt the governor had lost his way a bit after his father, Mario, had passed away. According to the adviser, Andrew had become less aware of how he treated other people, and Chris had supplanted Mario as a ballast for him in that regard.

Whatever Chris said that day worked. Andrew ultimately backed down and delivered a significantly more muted rebuttal to the allegations, essentially denying them and asking New Yorkers to allow the outside investigation to conclude. It was one of the last calls we'd have as a group—no more allegations came out publicly, the AG investigation had started to move quickly, and truthfully most of us felt pretty burned by the whole situation. There were the accusations, which had gotten increasingly more troubling: none of us were okay with enabling anyone who could have done such things.

Then, of course, there was the reckoning each of us had to have with ourselves. One false allegation, even two, you might give someone the benefit of the doubt. But with three or more, it's a pattern. And short of a vast conspiracy, there is likely an element of truth to that pattern. Had we been *had* this entire time? Had the Governor looked each of us in the eye and lied to us? Is it possible that the Governor himself believed his lies?

People have asked me why I stuck around and continued to advise him, even after I started to have doubts about his conduct and the things he was telling us. It's not like I was totally blind to the fact that political figures could lie or let me down. I'd dated Eliot Spitzer, for god's sake. I'd seen the worst of politics up close. But I'd also seen the best of it. There was never a day that I showed up to work for Pete or was on a call with him when I doubted his truthfulness or sincerity. Pete had redeemed my faith in the political process and reaffirmed why I'd chosen this line of work in the first place. I *wanted* to believe Cuomo, I *had* to. To me, the other option was unfathomable: that so much of what I'd done in politics, everything I'd done for Cuomo, was in vain. That I was just another sucker, another cog in a nihilistic machine.

There was also the fog of war that came with being in the middle of a crisis of that magnitude. Every day, it felt like there was incoming that needed an immediate response—allegations of misconduct, calls for him to resign, editorials scorching him. There was no time for deep analysis or self-reflection; it was just about getting through another news cycle and keeping the governor from succumbing to his worst instincts. If we weren't there, what crazy or offensive things would he say? We talked him down from batshit crazy statements and op-eds that would have ruined him on the spot. And, by this time, it wasn't just his name on the line—it was all of ours as well. The thought process was "*how* can we get him through this," not "*should* we help him get through this?" I should've ruminated on the second question more.

Fool me twice, shame on me.

I didn't hear from the governor for a number of months. He ran with the Northam playbook in the meantime. Every week that spring and summer, you could find him holding a press conference with Black clergy, community leaders, and elected officials. At one memorable event, Charlie Rangel—the ninety-year-old former Harlem Congressman with a legendary rap sheet of ethics violations—waved his cane at a podium and invoked Jesus Christ in defense of the governor. He told people to "back off until you've got some facts"—all while Cuomo, standing by Rangel's side, barely hid the large grin under his black cloth mask.

The governor reached out once in June, when he received a briefing of a statewide survey that his pollster had conducted (it found that Black and women voters were the most likely to give him the benefit of the doubt). Then he reached out again in July after he'd been interviewed by investigators for the Attorney General report. He called a small group of us into his office and was positively giddy about the interview: "Good news, gang," he told us. "I sat down with Joon Kim and Anne Clark [the AG's investigators] and there's going to be nothing new in the report. It will just be a rehash of everything from the spring."

He wanted to map out a prebuttal to the AG report, which he assured us would be an underwhelming document. On the docket? A letter from his attorneys contesting the objectivity of the investigators and a video, whose script he'd written out himself—a script that clocked in at over twelve minutes. It was vintage "Bad Andrew." The letter, which I suspect wasn't drafted by his lawyers, included ad hominem attacks against the AG's investigators and the previous US attorney from the Southern District of New York. "Diary of a Psychopath" is how I described it to another adviser. The video script, if possible, was even worse. It included a long section that he intended as a photo montage, where he'd show photo after photo of him kissing people of all ages, races, sexualities, and genders on the face—unaware of how tone deaf and even comical it would seem. "Governor, you should not do this," I told him. "Unless you want this video to be mocked and replayed on the late-night shows." There wasn't an adviser in disagreement. The difference now, however, was that Chris, the governor's brother, had fully extricated himself from all conversations regarding the governor. There was no one left who could close the deal with him.

In the end, none of it mattered. The AG called a last-minute nine A.M. Tuesday press conference on August 3, where she released the findings of the report. As expected, it addressed the allegations that had already been public. But it didn't stop there. My stomach dropped when I read the new bombshell finding in the report from a thirty-year-old female state trooper who'd served on the governor's security detail for the last two years. She told investigators that he had touched her inappropriately on repeated occasions and made comments of a wildly inappropriate sexual nature to her on the job—a job where she was tasked with protecting his life with hers.

Once again, he'd looked us all in the eyes and lied. Once again, he denied the charges and wanted to fight them. The difference this time was that no one around him believed him anymore, myself included. He called each of us individually to ask our opinions, seeking a sympathetic ear or some way out of the situation he found himself in. He didn't find one. The sole exception was former President Bill Clinton, who told him that he needed to go out and address the people of New York directly: to state that his fate was in their hands, not the politicians'. The consensus—among advisers, at least—was that unless Clinton, with his legendary political skills, was willing to do the mea culpa himself, it would do more harm than good.

For me, the AG's report was the last in the line of crushing blows. I'd been willing to overlook Cuomo's rough edges and obvious flaws—he'd done so much good in his eleven years as governor, and I'd seen plenty of the warm, caring side of him. But everything about the last several months had made me question my sanity and judgment. It made me wonder why I'd committed my life to a profession that was seemingly dominated by narcissists and liars. Every high I'd had had been matched by an even bigger low.

Within days, the man who had dominated Albany for the last eleven years, and had been floated as the next coming of the Democratic Party, announced his resignation.

"Fool me three times, shame on both of us."—Stephen King.

Cuomo's resignation wasn't the end of the whole affair. *Not even close.* It triggered a tsunami that destroyed everything in its path. The collateral damage was almost unfathomable.

There was Time's Up—the Hollywood-backed, post-#MeToo nonprofit that fully dissolved within weeks of the AG report's release (its CEO and a board member had given behind-the-scenes advice to Cuomo). Then there were the former aides who were forced out of high-profile gigs—the president of the Human Rights Campaign, the largest LGBTQ+ advocacy group in the country, and the chancellor of the 420,000-student State University of New York, among others. Chris was fired from hosting the top-rated show on CNN. And just two months later, CNN Worldwide's president, Jeff Zucker, was pushed out of his role, in part, for crossing journalistic lines in his own relationship with Andrew.

Those were the firings and resignations that made the headlines.

But there was also the untold story of the dozens and dozens of people who didn't merit an audience with the *New York Times*— the lower level staffers who worked in policy, intergovernmental relations, operations, and different state agencies. They put up with a tough, sometimes toxic workplace because working in the office of the New York governor was a badge of honor. It was something they could be proud of and it would obviously lead all of their resumes.

Overnight, that was taken from them. Some lost their jobs, others were denied opportunities that they should've been given—all because of their association with Cuomo. But it wasn't just what they lost, it was all the things they'd never get back—the years they'd wasted in his office, the life events they'd missed, and the personal relationships they'd strained due to the demanding nature of working for him.

Once again, I became a target in the press for my proximity to a man acting badly. On my worst days, I convinced myself that I'd reversed all of the professional gains I'd made in the last few years. Luckily, that wasn't the case. But it was excruciating to go through.

Say what you will about Andrew, but he died as he lived: with zero regard for the people around him and the impact his actions would have on them.

FOURTEEN

Curtain Call

Watching Cuomo implode felt like a death of sorts. In many ways, it was much more painful than any campaign loss I'd been a part of. I remember thinking to myself, naively, *How much worse could it get than this.* I'd soon get the answer—one that would put it all in perspective.

A week after Cuomo announced his resignation, I got a late-night call from my mom. My dad was found unresponsive in his room at his assisted living facility. He was rushed to White Plains Hospital and immediately intubated in the intensive care unit and placed in a medically induced coma. "Come as soon as you can in the morning," my mom told me. "This could be the end."

For the next sixteen agonizing days, my mom and I spent every minute we could by my dad's bedside. Watching my mom process losing her partner of forty-nine years was almost as painful as watching my dad's health decline. Disposition-wise, she's about as different from me as you could imagine. She's extroverted and sunny—someone who always sticks out in a crowd due to her five-foot-ten frame and bright blond hair. She's fearless and adventurous; she makes fast friends of strangers and almost never cries. For the first time in my life, I saw that cheery, strong veneer crack. She cried every morning on the drive from Bronxville to White Plains. In the hospital room, she rarely left my dad's side. I'd look over to see her rocking back and forth: "Tommy. Tommy. What am I going to do without you?"

I did my best to take some of the weight off her shoulders, relieving her at my dad's bedside every few hours. I read news stories aloud to him—everything from coverage of the United States' disastrous withdrawal from Afghanistan to clips about his beloved Boston Red Sox and New York Jets. I played his favorite music on a Bluetooth speaker—Dvorak's *New World Symphony*, the Eagles, and the score to Martin Scorsese's documentary *The Last Waltz*—a soundtrack he'd relied on to lull Angus and me to sleep when we were babies. I recounted to him my favorite memories from growing up—the times he'd let Angus and me run wild around the World Trade Center when he'd work from the office on weekends. The Halloween when he didn't have a costume, but greeted trick-or-treaters in character as the 1996 Republican nominee for president: "Bob Dole thinks you should take a Kit Kat bar. Bob Dole wants you to take these coins for your Unicef box." (Bob Dole, a decorated veteran of World War II and former majority leader in the Senate, spoke often of himself the third person.) I confided how I'd kept our secrets from our time as roommates in DC—the Sapporos and healthy pours of scotch we'd throw back on Sunday nights—from Mom (she wasn't a big drinker). I came clean on some white lies I'd told him in high school: yes, I hadn't actually scratched the passenger's side of his car and snapped off the side-view mirror by backing out of our driveway (Angus and I had, in reality, badly miscalculated the width of an alley we'd been driving through one night). And every day, I made sure to tell him how much I loved him. How lucky I'd been to have the best dad in the world. How I never would've been able to go into the unlikely world of politics and continue in it without his belief in me.

Five days into his stay in the ICU, the doctors took him off his ventilator. They asked my older brother, Toby, and me to leave for the extubation—it wouldn't be pretty. When the nurse came to find us in the waiting room, Toby and I looked up at him, expecting the worst—that our dad had passed away. Except he hadn't. He held on for eleven more days. He never fully regained consciousness—occasionally, when I'd talk to him, he'd open his eyes. Sometimes, a single tear would run down his face.

He passed away at three P.M. on a Thursday afternoon—his left hand in my mom's, his right hand in mine, with Toby sitting at the foot of his bed, his hands on my dad's leg. We told him over and over how much we loved him. That it was okay for him to let go and move on. That one day we would all meet again.

After he drew his last ragged breaths and the life left his body, I walked out of the hospital in a trance. I went to the corner convenience store and bought a half dozen Fireball minis. I downed four sitting on a bench outside, waiting for my mom and Toby to come out so that we could get as far from the hospital as possible.

I'd experienced grief before in my life, on many different levels. Grandparents or friends who'd died. Romantic relationships that had ended poorly. Crushing campaign defeats. A governor's resignation. But nothing held a candle to this.

A few weeks after his death, I got a letter from President Biden, someone who knows a thing or two about loss. It ended: "Though the grieving process never truly ends, I promise you the day will come when your father's memory brings a smile to your lips before it brings a tear to your eye. My prayer is that this day comes sooner rather than later."

That day hasn't come yet. But I've seen glimpses of it.

Since his passing, I've tried to stay close to my dad by doing some of the things he loved. Chief among them—going to every Jets home game (my dad had adopted them as his NFL team when he moved to New Jersey in 1955 to attend Princeton). For years, he'd dragged me to the Meadowlands, and then to MetLife Stadium, wearing—without exception—his Joe Namath jersey. Namath, famously, had led the Jets to a historic Super Bowl upset victory in the 1968 season. In the fifty-plus years since, however, the team hadn't set foot in the Big Game. Still, my dad stuck by them through thick and thin—inspired seasons and truly awful ones. To him, every week brought with it an exhilarating possibility—that any given Sunday, no matter the odds, no matter the formidability of the opposition, the Jets could emerge victorious. The team may have seemed like a lost cause to even the most casual football fan, but never to my dad. Every game, he was there with his binoculars—rooting them on, convinced that they were just one strategic call away from greatness. Every season was a chance for them to start anew—*this* was the year they would win the Super Bowl.

As clichéd as it might sound, I felt my dad next to me during the eighth game of the 2021 season, seven weeks after he'd passed away. I was at the Jets' showdown with the heavily favored Bengals, expecting a massacre. All of their stars were injured. The second-string quarterback—someone even diehard football fans like me had never heard of—made his first career start. Only half the stadium was full because no one expected it to be a game at all. And then, Mike White—the obscure quarterback—threw for over four hundred yards and three touchdowns, clocking the best starting performance for any QB in the NFL since 1950. The Jets stunned the league with their win; there was not an oddsmaker who would've recommended betting a dime on them. As the stadium cleared out, I stayed behind with my companion looking out across the empty field and eighty-two thousand empty seats: "God, I wish my dad had seen this game." Eventually, two burly Jersey ushers told us politely to beat it: "But take those beers with you," they added. "You know those cost twelve dollars here."

We have a saying in this business: every year spent on the campaign trail is the equivalent of a dog year. By that measure, by the time I turned thirty-nine, I'd spent 130 years in the trenches of American politics. And trust me, I felt every single ache and pain that came with them. Seventeen years in politics have taken their toll on me personally and professionally. The last year, the most. It's tempting to think I could simply walk away from the business.

Politics, like football, involves an element of faith. You can have a good year—a great candidate, a winning campaign, accolades for being a strategic genius—and feel like you're on top of the world. Then, you turn around and get completely shellacked in your next race, and you question why you're even in the business to start with. Plenty of people I've worked with have packed up their bags after a brutal loss and left politics behind completely.

But if there's one thing my dad instilled in me, it's the value of belief and loyalty. (Earned loyalty, not blind loyalty—as I learned with the Cuomo situation.) Even after the most brutal, lopsided loss imaginable, you can pick yourself up and find another worthwhile candidate or cause or campaign that can make a difference in people's lives. You have to.

The L's I've taken don't define politics for me. I've learned that around the corner of every disappointment, there is possibility; that for every politician who lets you down—even after you put your faith in them—there is a new, fresh face who can redeem your belief in the process. I still believe in the power of politics to improve people's lives. I still believe that there is hope for the future—that every day, week, and year there are future leaders coming out of the woodwork who have the potential to change the world, if we give them a chance. I've been blessed to have encountered some of them in life, and I know that I will again.

I'm my father's daughter. I'll always believe in the possibility of any given Tuesday. Whether on the field or in the stands, I'll always show up to the game.

Acknowledgments

Of all the things I thought I'd never do in life, writing a book ranked near the top of the list—somewhere in between dyeing my hair blond and running for elected office. It was an eighteen-month labor of love that wouldn't have been possible without the encouragement and help of so many people.

My mother, Adrienne Smith, served as a ballast for me throughout the process, and I'm grateful for the love that she and my siblings, Angus, Toby, and Ashley have shown me across the last thirty-eight years. I needed it especially during the writing of this book.

This project never would have come to fruition without my friend and guardian angel Matt Hiltzik, who raised the prospect with me after the 2020 presidential primaries and didn't let it rest until he'd helped me find an agent, Sloan Harris, who appreciated my unique voice. Sloan read every draft proposal, chapter, and manuscript of mine, helping turn a girl who'd never written anything longer than a term paper into a published author. At HarperCollins, my editors Jennifer Barth and—later—Jonathan Jao helped me cull every experience I've had in politics into a

Acknowledgments

tightly written three-hundred-page memoir. I'm appreciative for the kind words that David Axelrod, Karen Tumulty, and Bakari Sellers offered for it.

I make note of my professional mentors in the book, but I've been incredibly lucky to be guided by David Axelrod, Stephanie Cutter, and Ben LaBolt over the last decade. Whenever I needed professional advice, they were there for me and they helped me navigate some truly tricky times.

To faithfully recreate the events in each chapter, I called upon friends and coworkers from over the years who graciously lent me their time. In alphabetical order they are: Sarah Berlenbach, Kurt Bardella, Abbey Collins, Sean Darcy, Melissa DeRosa, Jonathan Davis, Mo Elleithee, Dave Hamrick, Miriam Hess, Bill Hyers, Danny Kanner, Charlie King, Dani Lever, Rob Lockwood, Janos Marton, Patrick McKenna, Tim Miller, Haley Morris, Steve Neuman, Colm O'Comartun, Justin Paschal, Jefrey Pollock, Erick Sanchez, Tait Sye, Phil Walzak, and Sharon Yang.

I want to give special recognition to the members of Team Pete who helped hugely in this effort: Katie Connolly, John Del Cecato, Kevin Donohoe, Larry Grisolano, Ben Halle, Michael Halle, Chuck Kennedy, Brendan McPhillips, Hari Sevugan, Mike Schmuhl, and Maya Shankar. We created the best communications operation in presidential campaign history—Team Dreadnought—that served as my unofficial fact-checking chain for this book. Thank you to Ro Applewhaite, Manuel Bonder, Matt Corridoni, Zev Karlin-Neuman, Andrew Mamo, Chris Meagher, Marisol Samoyoa, Sean Savett, Nina Smith, and Tess Whittlesey. And, of course, I am forever grateful to Pete—he gave me my dream job and helped me believe in politics (and myself) again.

Lastly, there are the friends who provided invaluable emotional support when I needed it most. They read my drafts, suggested edits, and talked me off the ledge too many times to count. My

Acknowledgments

Dartmouth roommates (and best friends in the world)—Nina Edelman, Kassidee Kipp, and Margot Langsdorf—were stuck in upstate New York with me for the first two miserable months of producing a manuscript and miraculously stayed friends with me. Tammy Haddad helped me navigate the process from inception to blurbs, offering the steady guidance I needed in moments of panic. Jeff Smith, the best ex-boyfriend ever to walk the earth, spent countless hours with me on the phone and over text offering feedback. Eric Koch read more drafts of chapters and manuscripts than anyone except Sloan and was my MVP who helped me craft the most sensitive sections of the book. And, finally, I owe a huge thank-you to Amanda Konstam, who—despite not being a professional football fan—first came up with the title *Any Given Tuesday*.

Writing a memoir can be a deeply painful and raw experience at times. Most of all, though, it reminded me how lucky I am to have lived out my dreams with the most amazing people by my side every step of the way.

About the Author

LIS SMITH is a veteran of twenty political campaigns. She has extensive experience in public affairs and media relations at the local, state, and national levels. She most recently worked as a senior adviser in communications to presidential candidate Pete Buttigieg. Prior to working on Buttigieg's campaign, she'd worked for everyone from former president Barack Obama to Senator Claire McCaskill and Governors Terry McAuliffe, Ted Strickland, and Martin O'Malley. She's served as an on-air commentator for major television networks and has had opinion pieces published in the *New York Times* and *Vanity Fair*. She lives in New York City.